Lecture Notes in Computer Science 12445

More information about this series at http://www.springer.com/series/7412

Tanveer Syeda-Mahmood ·
Klaus Drechsler et al. (Eds.)

Multimodal Learning for Clinical Decision Support and Clinical Image-Based Procedures

10th International Workshop, ML-CDS 2020
and 9th International Workshop, CLIP 2020
Held in Conjunction with MICCAI 2020
Lima, Peru, October 4–8, 2020
Proceedings

 Springer

Editors
Tanveer Syeda-Mahmood 🄳
IBM Almaden Research Center
San Jose, CA, USA

Klaus Drechsler
Aachen University of Applied Sciences
Aachen, Germany

Additional Workshop Editors *see next page*

ISSN 0302-9743 ISSN 1611-3349 (electronic)
Lecture Notes in Computer Science
ISBN 978-3-030-60945-0 ISBN 978-3-030-60946-7 (eBook)
https://doi.org/10.1007/978-3-030-60946-7

LNCS Sublibrary: SL6 – Image Processing, Computer Vision, Pattern Recognition, and Graphics

This Springer imprint is published by the registered company Springer Nature Switzerland AG
The registered company address is: Gewerbestrasse 11, 6330 Cham, Switzerland

Additional Workshop Editors

ML-CDS 2020 Editors

Hayit Greenspan
Tel Aviv University
Ramat Aviv, Israel

Anant Madabhushi
Case Western Reserve University
Cleveland, OH, USA

Alexandros Karargyris
IBM Almaden Research Center
San Jose, CA, USA

CLIP 2020 Editors

Cristina Oyarzun Laura
Fraunhofer-Institute
for Computer Graphics Research (IGD)
Darmstadt, Germany

Marius George Linguraru
Children's National Hospital
Washington, DC, USA

Stefan Wesarg
Fraunhofer-Institute for Computer Graphics
Research (IGD)
Darmstadt, Germany

Raj Shekhar
Children's National Hospital
Washington, DC, USA

Marius Erdt
Fraunhofer Singapore
Singapore, Singapore

Miguel Ángel González Ballester
Universitat Pompeu Fabra
Barcelona, Spain

Preface ML-CDS 2020

On behalf of the organizing committee, we welcome you to the 10th Workshop on Multimodal Learning for Clinical Decision Support. The goal of these series of workshops is to bring together researchers in medical imaging, medical image retrieval, data mining, text retrieval, and machine learning/AI communities to discuss new techniques of multimodal mining/retrieval and their use in clinical decision support. Although the title of the workshop has changed slightly over the years, the common theme preserved is the notion of clinical decision support and the need for multimodal analysis. The previous seven workshops on this topic have been well-received at MICCAI, specifically, in Shenzen (2019), Granada (2018), Quebec City (2017), Athens (2016), Munich (2015), Nagoya (2013), Nice(2012), Toronto (2011), and London (2009).

Continuing on the momentum built by these workshops, our focus remains on multimodal learning. As has been the norm with these workshops, the papers were submitted in 8 page double-blind format and were accepted after review. The workshop continued to stay with an oral format for all the presentations. The day ended with a lively panel composed of more doctors, medical imaging researchers, and industry experts. This year we also invited researchers to participate in a tubes and lines detection challenge within the program. Finally, as has been our tradition, our invited speakers were from the clinician world. This year, we highlighted progress in digital pathology with Dr. Michael Feldman from the University of Pennsylvania as our invited speaker.

With less than 5% of medical image analysis techniques translating to clinical practice, workshops on this topic have helped raise the awareness of our field to clinical practitioners. The approach taken in the workshop is to scale it to large collections of patient data exposing interesting issues of multimodal learning and its specific use in clinical decision support by practicing physicians. With the introduction of intelligent browsing and summarization methods, we hope to also address the ease-of-use in conveying derived information to clinicians to aid their adoption. Finally, the ultimate impact of these methods can be judged when they begin to affect treatment planning in clinical practice.

We hope you enjoyed the program we have assembled and actively participated in the discussion on the topics of the papers and the panel.

October 2020

Tanveer Syeda-Mahmood
Hayit Greenspan
Anant Madabhushi
Alexandros Karargyris

Organization

Program Chairs

Tanveer Syeda-Mahmood	IBM Research, USA
Alexandros Karargyris	IBM Research, USA
Hayit Greenspan	Tel-Aviv University, Israel
Anant Madabhushi	Case Western Reserve University, USA

Program Committee

Amir Amini	University of Louisville, USA
Sameer Antani	National Library of Medicine, USA
Rivka Colen	MD Andersen Research Center, USA
Keyvan Farahani	National Cancer Institute, USA
Alejandro Frangi	The University of Sheffield, UK
Guido Gerig	The University of Utah, USA
David Gutman	Emory University, USA
Allan Halpern	Memorial Sloan-Kettering Research Center, USA
Ghassan Hamarneh	Simon Fraser University, Canada
Jayshree Kalpathy-Kramer	Massachusetts General Hospital, USA
Ron Kikinis	Harvard University, USA
Georg Langs	Medical University of Vienna, Austria
Robert Lundstrom	Kaiser Permanente, USA
B. Manjunath	University of California, Santa Barbara, USA
Dimitris Metaxas	Rutgers, USA
Nikos Paragios	École Centrale Paris, France
Daniel Racoceanu	National University of Singapore, Singapore
Eduardo Romero	Universidad Nacional de Colombia, Colombia
Daniel Rubin	Stanford University, USA
Russ Taylor	Johns Hopkins University, USA
Agma Traina	University of São Paulo, Brazil
Max Viergewer	Utrecht University, The Netherlands
Sean Zhou	Siemens Corporate Research, USA

Preface CLIP 2020

On October 4, 2020, the 9th International Workshop on Clinical Image-based Procedures: From Planning to Intervention (CLIP 2020), was held in conjunction with the 23rd International Conference on Medical Image Computing and Computer Assisted Intervention (MICCAI 2020). Due to the COVID-19 pandemic, the workshop was held as an online-only meeting to contribute to slowing down the spread of the virus. Despite the challenges involved, we have continued to build on what we have successfully practiced over the past eight years: providing a platform for the dissemination of clinically tested, state-of-the-art methods for image-based planning, monitoring, and evaluation of medical procedures.

A major focus of CLIP 2020 was on the creation of holistic patient models to better understand the need of the individual patient and thus provide better diagnoses and therapies. In this context, it is becoming increasingly important to not only base decisions on image data alone, but to combine these with non-image data, such as 'omics' data, electronic medical records, electroencephalograms, and others. This approach offers exciting opportunities to research. CLIP provides a platform to present and discuss these developments and work, centered on specific clinical applications already in use and evaluated by clinical users.

In 2020, CLIP accepted nine original manuscripts from all over the world for oral presentation at the online event. Each of the manuscripts underwent a single-blind peer review by two members of the Program Committee, all of them prestigious experts in the field of medical image analysis and clinical translations of technology. We would like to thank our Program Committee for its invaluable contributions and continuous support of CLIP over the years. It is not always easy to find the time to support our workshop given full schedules and challenges due to the ongoing pandemic, and we are very grateful to all our members because CLIP 2020 would not have been possible without them. We would also like to thank all the authors for their high-quality contributions this year as well as their efforts to make CLIP 2020 a success. Finally, we would like to thank all MICCAI 2020 organizers for supporting the organization of CLIP 2020.

October 2020

Klaus Drechsler
Marius George Linguraru
Cristina Oyarzun Laura
Raj Shekhar
Stefan Wesarg
Miguel Ángel González Ballester
Marius Erdt

Organization

Organizing Committee

Klaus Drechsler	Aachen University of Applied Sciences, Germany
Marius Erdt	Fraunhofer Singapore, Singapore
Miguel González Ballester	ICREA, Universitat Pompeu Fabra, Spain
Marius George Linguraru	Children's National Hospital, USA
Cristina Oyarzun Laura	Fraunhofer IGD, Germany
Raj Shekhar	Children's National Hospital, USA
Stefan Wesarg	Fraunhofer IGD, Germany

Program Committee

Yufei Chen	Tongji University, China
Jan Egger	TU Graz, Austria
Chaoqun Dong	Fraunhofer Singapore, Singapore
Katarzyna Heryan	AGH University of Science and Technology, Poland
Yogesh Karpate	Gauss and Riemann Scientific, India
Roman Martel	Fraunhofer Singapore, Singapore
Luís Rüger Sacco	Fraunhofer Singapore, Singapore
Xingzi Zhang	Fraunhofer Singapore, Singapore
Stephan Zidowitz	Fraunhofer MEVIS, Germany

Contents

CLIP 2020

Optimal Targeting Visualizations for Surgical Navigation
of Iliosacral Screws... 3
    Prashant U. Pandey, Pierre Guy, Kelly A. Lefaivre,
    and Antony J. Hodgson

Prediction of Type II Diabetes Onset with Computed Tomography
and Electronic Medical Records 13
    Yucheng Tang, Riqiang Gao, Ho Hin Lee, Quinn Stanton Wells,
    Ashley Spann, James G. Terry, John J. Carr, Yuankai Huo,
    Shunxing Bao, and Bennett A. Landman

A Radiomics-Based Machine Learning Approach to Assess Collateral
Circulation in Ischemic Stroke on Non-contrast Computed Tomography 24
    Mumu Aktar, Yiming Xiao, Donatella Tampieri, Hassan Rivaz,
    and Marta Kersten-Oertel

Image-Based Subthalamic Nucleus Segmentation for Deep Brain Surgery
with Electrophysiology Aided Refinement 34
    Igor Varga, Eduard Bakstein, Greydon Gilmore, and Daniel Novak

3D Slicer Craniomaxillofacial Modules Support Patient-Specific
Decision-Making for Personalized Healthcare in Dental Research 44
    Jonas Bianchi, Beatriz Paniagua, Antonio Carlos De Oliveira Ruellas,
    Jean-Christophe Fillion-Robin, Juan C. Prietro,
    João Roberto Gonçalves, James Hoctor, Marília Yatabe, Martin Styner,
    TengFei Li, Marcela Lima Gurgel, Cauby Maia Chaves Junior,
    Camila Massaro, Daniela Gamba Garib, Lorena Vilanova,
    Jose Fernando Castanha Henriques, Aron Aliaga-Del Castillo,
    Guilherme Janson, Laura R. Iwasaki, Jeffrey C. Nickel,
    Karine Evangelista, and Lucia Cevidanes

Learning Representations of Endoscopic Videos to Detect Tool Presence
Without Supervision 54
    David Z. Li, Masaru Ishii, Russell H. Taylor, Gregory D. Hager,
    and Ayushi Sinha

Single-Shot Deep Volumetric Regression for Mobile Medical
Augmented Reality 64
    Florian Karner, Christina Gsaxner, Antonio Pepe, Jianning Li,
    Philipp Fleck, Clemens Arth, Jürgen Wallner, and Jan Egger

A Baseline Approach for AutoImplant: The MICCAI 2020 Cranial Implant
Design Challenge   75
    Jianning Li, Antonio Pepe, Christina Gsaxner, Gord von Campe,
    and Jan Egger

Adversarial Prediction of Radiotherapy Treatment Machine Parameters   85
    Lyndon Hibbard

ML-CDS 2020

Soft Tissue Sarcoma Co-segmentation in Combined MRI
and PET/CT Data   97
    Theresa Neubauer, Maria Wimmer, Astrid Berg, David Major,
    Dimitrios Lenis, Thomas Beyer, Jelena Saponjski, and Katja Bühler

Towards Automated Diagnosis with Attentive Multi-modal Learning Using
Electronic Health Records and Chest X-Rays   106
    Tom van Sonsbeek and Marcel Worring

LUCAS: LUng CAncer Screening with Multimodal Biomarkers   115
    Laura Daza, Angela Castillo, María Escobar, Sergio Valencia,
    Bibiana Pinzón, and Pablo Arbeláez

Automatic Breast Lesion Classification by Joint Neural Analysis
of Mammography and Ultrasound   125
    Gavriel Habib, Nahum Kiryati, Miri Sklair-Levy, Anat Shalmon,
    Osnat Halshtok Neiman, Renata Faermann Weidenfeld, Yael Yagil,
    Eli Konen, and Arnaldo Mayer

Author Index   137

CLIP 2020

Optimal Targeting Visualizations for Surgical Navigation of Iliosacral Screws

Prashant U. Pandey[1]($\boxtimes$), Pierre Guy[2], Kelly A. Lefaivre[2], and Antony J. Hodgson[3]

[1] Biomedical Engineering, University of British Columbia, Vancouver, Canada
prashant@ece.ubc.ca
[2] Department of Orthopaedics, University of British Columbia, Vancouver, Canada
[3] Mechanical Engineering, University of British Columbia, Vancouver, Canada

Abstract. Surgical navigation can be used for complex orthopaedic procedures, such as iliosacral screw (ISS) fixations, to achieve accurate results. Although studies have documented the impact of navigation on surgical outcomes, no works to date have quantified the effects of how information regarding the surgical navigation scene is displayed to surgeons on conventional monitors. However, visualizing information in different ways can have a measurable effect on both accuracy and time required to perform the navigated task. We conducted a user study to investigate which visualization techniques helped non-expert users effectively navigate a surgical tool to screw targets. We proposed a series of 2D and 3D visualizations with varying representations of the targets and tool, and measured the time required and accuracy of each participant to perform the task. We found that a bullseye view and a 'target-fixed with bullseye' view allowed users to most efficiently complete simulated pelvic screw targeting, with mean accuracies of 0.47 mm and 0.73° and 0.66 mm and 0.50° respectively. Furthermore, our study found that 3D visualizations on their own are significantly less accurate and efficient, and that the orientation of the virtual surgical scene must match the user's perspective of the physical scene to prevent unnecessary time mentally reconciling the two domains. These results can more systematically guide the development of visualizations in a surgical navigation system for screw insertions.

Keywords: Surgical navigation · Iliosacral screw insertion · Human computer interaction

1 Introduction

Surgically repairing a pelvic fracture is a complex procedure which relies heavily on intraoperative X-ray imaging, thus exposing surgical staff to high doses of

Electronic supplementary material The online version of this chapter (https://doi.org/10.1007/978-3-030-60946-7_1) contains supplementary material, which is available to authorized users.

T. Syeda-Mahmood et al. (Eds.): ML-CDS 2020/CLIP 2020, LNCS 12445, pp. 3–12, 2020.
https://doi.org/10.1007/978-3-030-60946-7_1

ionizing radiation [5]. Furthermore, surgeons are required to perform 'mental gymnastics' during the surgery by referring to multiple 2D or 3D X-ray and CT images to determine the best trajectory to use for inserting screws into the patient. Manipulating these images or acquiring new images is not straightforward and can be a time-consuming process which increases operation duration, and also leads to inaccuracies [6,7]. Surgical navigation is a potential solution, using tracking sensors combined with intraoperative imaging technologies [8–10]. Navigation allows the surgeon to visualize the position of surgical tools relative to the surgical site of interest which is typically extracted from a preoperative CT scan. However, this data can be presented in many different formats, such as 2D images or 3D models, combined with arbitrary viewing angles and planes, on a conventional monitor. This information can be potentially overwhelming and confusing.

Although there have been recent developments in using augmented reality (AR) devices for orthopaedic surgical navigation and visualization, such as head mounted displays (HMDs) [2,3] and mobile displays [4], these devices have not yet been widely adopted in operating rooms. In particular, using commercially-available HMDs is challenging due to their size and weight and limited field-of-view, and they can be disruptive to the surgical work- flow [1]. In contrast, 2D monitors are widely-used in operating rooms for many tasks such as medical image visualization and displaying patient vital signs.

To the best of our knowledge, there are no prior studies that have quantified how visualizing the surgical navigation scene on a conventional monitor affects targeting performance for orthopaedic surgical implants. It has been shown, however, that the display of information can have a measurable effect on both accuracy and required time to perform general non-domain specific targeting tasks [11]. Furthermore, it has been shown that 3D instead of 2D representations helped medical residents and surgeons better localize tumors preoperatively for neurosurgical tumor resection when assessed through a multiple choice test [12]. We therefore expect to find that surgical targeting performance is similarly dependent on visualizations, and believe that this should be quantitatively measured in the context of orthopaedic procedures. Therefore, we designed a user study to determine the optimal visualizations for performing navigated pelvic screw targeting, which is a common orthopaedic surgical task during a complex surgery. Because using a targeting display is largely a motor-coordinative task, performance when using it should be relatively independent of medical or surgical expertise. For this pilot study, therefore, we felt it was sufficient to recruit study participants without explicit medical or surgical training and no more than moderate anatomical knowledge. The findings of this study can be used to more systematically design visualization components of a navigation system.

2   Methods

We conducted a user study to investigate which visualization techniques helped users effectively navigate a surgical tool to align with a screw axis target – a core task during pelvic fracture fixations. Because our primary focus was on evaluating the influence of targeting visualizations on a standardized motor-coordinative task that primarily requires the user to align a tool axis with a target axis, we limited the study's scope to only the initial drill targeting/alignment phase and not the subsequent hole-drilling phase. We simulated the surgical procedure using a dry pelvis model and optical tracking. We designed and evaluated multiple 2D and 3D visualizations to test how changes in perspectives, alignment, and relative motion between the tool and target affected targeting performance. We measured the accuracy and time required by each participant to perform the task while using each of the different visualizations.

Materials

We acquired a CT scan of a radio-opaque pelvis bone model (Sawbones, WA, USA), and placed 10 virtual cylindrical targets (diameter: 7.3 mm and varying lengths) within the CT image using the open-source Slicer software [13]. Only one target was presented at a time to the user, resulting in 10 unique targeting scenarios. Each cylindrical target was placed to mimic a unilateral iliosacral screw (ISS) inserted into the right of the S1 body in the sacrum. To simulate a navigated surgery, we tracked the pelvis model and a surgical wire guide tool using a Polaris Vega optical camera and rigid body markers (Northern Digital, ON, Canada). We calibrated the wire guide to define its tip and axis relative to the attached marker [14], and subsequently performed manual point-based registration to align the pelvis to its CT image using embedded metal screws as the fiducials.

Visualizations

We designed one 3D and five 2D visualizations to represent the surgical scene, using Slicer and SlicerIGT [15]. The 2D visualizations were:

1. '**Bullseye**' - looking 'down' the screw channel, oriented such that the anatomical axes on the display were aligned with the physical pelvis
2. '**Rotated bullseye**' - bullseye view rotated 90° clockwise with respect to the physical pelvis
3. '**Target-fixed**' - two views showing the target's cross-sections along the screw axis in two orthogonal planes nominally aligned with axial and coronal planes. The target and CT slice were fixed in position and orientation while the tool motion was displayed
4. '**Tool-fixed (translation)**' - the same two orthogonal planes as in 'target-fixed', but in this mode the tool-tip's position was fixed in the centre of the views and the CT was resliced and translated during tool motion. The target was projected onto each resliced plane
5. '**Tool-fixed (translation and rotation)**' - similar to 4, but here the two orthogonal planes were defined by the tool's pose. These planes resliced the CT during tool motion, and the tool-tip's position and orientation were fixed in the center of the views, while the CT was rotated and translated during tool motion. This mode is also known as 'dynamic reslicing' in SlicerIGT's terminology [15].

The bullseye and 'tool-fixed (translation and rotation)' visualizations were chosen to mimic commonly used modes in commercial and open-source navigation frameworks. The 'rotated bullseye' view was designed to test how discrepancies in alignment between the physical and virtual model affect navigation performance, whereas the 'target-fixed' and 'tool-fixed (translation)' visualizations were designed to understand how changes in relative motion between the tool, image and target affect navigation performance.

The **3D visualization** was designed to display two volumetric renderings of the CT scan approximately representing the perspective of the user looking at the physical pelvis model. Each rendering was cropped by orthogonal planes defined by the tool's pose as in 5 above. We chose volumetric renderings instead of isosurface models so that both internal and external anatomical structures of the pelvic bone could be represented.

All views represented the tool as a cyan cylindrical model coupled with a blue trajectory line. Each screw target was represented as a yellow cylinder with a green mid-line axis. The 2D views also displayed the relevant CT slice at either the location of the target (for the bullseye, 'rotated bullseye', and 'target-fixed' views) or at the location of the tool (for both the 'tool-fixed' views). The tool and tool trajectory were projected onto the visualization plane for all methods, with the exception of the 'tool-fixed (translation and rotation)' and 3D views. Finally, a humanoid marker was provided in the bottom-right to help users understand the anatomical perspective of each view/plane.

Computations to achieve these visualizations were integrated into a Slicer module such that relevant views were automatically generated in real-time, given an input CT scan, screw target, and the tracked tool's linear transformation relative to the pelvis model. The bullseye, 'target-fixed' and 3D visualizations are illustrated in Fig. 1.

User Study Design

We recruited 10 novice participants who had no previous experience performing surgical fixations and using navigation during surgery (mean age: 26.2 ± 5.3 years, 5 males and 5 females). We asked each participant to navigate the tool to five targets for each of the visualizations, using only the information presented on the monitor (Fig. 2). The physical pelvis model was covered with a sheet so participants could not use direct visual information from the bone during the task, as in a minimally-invasive surgery. The order of visualizations was randomized for each participant. For each target, we recorded the tool trajectory relative to the target entry. Participants were responsible for indicating when they were satisfied that the tool was positioned as accurately as possible, at which point the trajectory recording was stopped. After each participant completed the 6 prescribed visualizations, we asked them which visualization they preferred most - with the exception of the two bullseye views. We then combined this preferred view with a bullseye, and asked them to navigate again to 5 random targets. In total, 35 trajectories were acquired per participant. The study was conducted under institutional ethics approval.

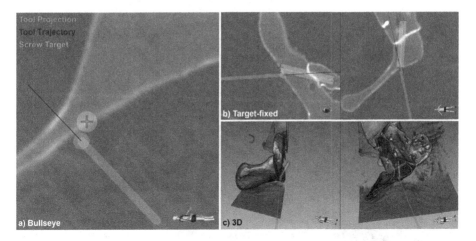

Fig. 1. Three of the six visualizations used in the study: a) bullseye, b) 'target-fixed' and c) 3D view. The other three visualizations differ only in orientation or in their dynamic representation during tool motion. The supplementary video illustrates the dynamic nature of all 6 visualizations and 3 additional bullseye-combination views.

Evaluation

To assess the performance of each visualization for navigated screw insertion, we measured the total time required to complete each target, and the time to reach a threshold accuracy (within 3.65 mm of target entry and 5° of the target axis - defined by the radius and length of a typical ISS). We calculated the distance error (ignoring the depth direction as participants were not able to place the tool tip on planes other than the screw entry), and rotational error at the final tool pose. We used a two-way analysis of variance (ANOVA) and the Tukey-Kramer post-hoc multiple comparisons test to infer statistical differences between the above metrics grouped by visualizations.

3   Results

Means and standard deviations of all measures are summarized in Table 1. Overall, the standard bullseye view had the lowest mean total required time for navigation (29.3 s) and this was statistically lower than the 'rotated bullseye' ($p < 0.01$), and 3D visualizations ($p < 10^{-6}$). The 'target-fixed with bullseye' combination had a mean time of 30.4 s, which was also statistically lower than both 'rotated bullseye' ($p < 0.05$) and 3D views ($p < 10^{-4}$). The 3D view had the longest total required time (mean: 74.3 s), which was statistically significant compared to the bullseye, 'tool-fixed (translation and rotation)', and 'target-fixed with bullseye' views (Fig. 3).

The bullseye and 'target-fixed with bullseye' views also had the lowest mean times to threshold of 13.5 s and 13.2 s respectively, which were both statistically lower than the 'target-fixed', 'tool-fixed (translation)', and 3D views.

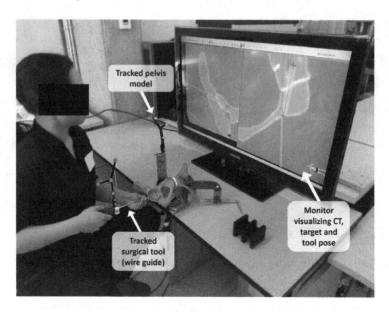

Fig. 2. User study setup. Each participant was asked to align the tracked surgical tool with the virtual targets using information from the visualizations displayed on the monitor. The pelvis model was covered with a sheet during the study, but is uncovered here for illustration.

Table 1. Means and standard deviations of each measure across all participants, grouped by visualization. **Green** indicates the mean was statistically lower than at least one of the other visualizations within a column, and **red** indicates the mean was statistically higher than at least one of the other visualizations within a column.

Visualization	Total time (s)	Time to 3.65mm and 5° (s)	Distance Error (mm)	Rotational Error (°)
Bullseye	**29.3 (19.1)**	**13.5 (14.7)**	**0.47 (0.42)**	**0.73 (1.71)**
Rotated Bullseye	**60.4 (40.0)**	**22.8 (15.0)**	0.60 (0.36)	0.50 (0.31)
Target-fixed	52.0 (49.4)	**27.5 (22.9)**	**0.90 (0.78)**	**1.12 (0.68)**
Tool-fixed (translation)	50.8 (24.1)	**27.2 (18.9)**	0.79 (0.48)	0.97 (0.55)
Tool-fixed (translation + rotation)	44.1 (23.4)	**22.0 (18.4)**	**0.73 (0.39)**	**0.96 (0.58)**
3D	**74.3 (73.3)**	**40.6 (36.3)**	**3.42 (2.25)**	**6.05 (5.47)**
Target-fixed + Bullseye	**30.4 (20.9)**	**13.2 (7.3)**	**0.66 (0.41)**	**0.58 (0.34)**
Tool-fixed (translation + rotation) + Bullseye	35.8 (15.3)	**15.5 (4.1)**	**0.56 (0.29)**	**0.41 (0.40)**
3D + Bullseye	56.5 (17.9)	15.0 (19.9)	**0.41 (0.33)**	**1.00 (0.24)**

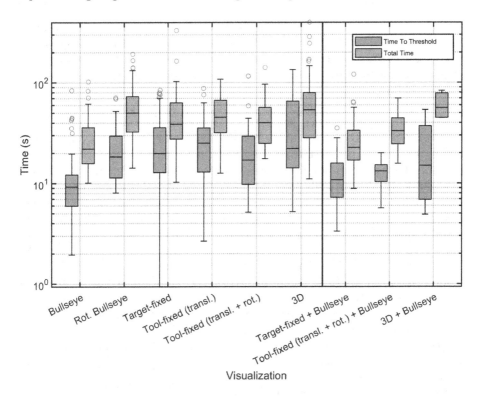

Fig. 3. Total required time and time to threshold (within 3.65 mm and 5°) grouped by visualization. The 3 groups to the right of the vertical line do not have equal samples to the groups on the left, as these visualizations were used based on each user's preferences. For each box, the middle horizontal line is the median, the height is the interquartile range, the whiskers extend to the minimum and maximum values excluding outliers. The y-axis is plotted on a logarithmic scale.

All 2D views and bullseye-combination views had statistically lower means for distance and rotational errors when compared to the 3D view alone, and there were no other statistical differences in accuracies. The mean distance and rotational errors ranged between 0.41–0.90 mm and between 0.4–1.1° respectively, for the 2D and bullseye-combination views.

Seven participants preferred the 'target-fixed' visualization, two the 'tool-fixed (translation and rotation)' view, and one the 3D view. We combined each participant's preference with a central bullseye view. In total, the participants generated 349 trajectories, which we grouped into the respective visualizations used while completing each targeting task.

4   Discussion and Conclusion

Our results indicate that the bullseye and 'target-fixed + bullseye' visualizations were the most effective ones for navigated screw targeting, given their statistically quicker times than other visualizations while maintaining equivalent accuracy. This is likely because the bullseye view was also the simplest: it allowed participants to align the tool with the target using only one perspective, unlike other visualizations which required users to align the tool using at least two perspectives.

Conversely, the 3D visualization was significantly worse than the others across all time and accuracy metrics, suggesting it is not an effective way to visualize navigation when used on its own. We believe this was because it was difficult for users to precisely visually interpret the pose of the tool with respect to a target given only 3D representations. However, when combined with a bullseye, the 3D visualization had statistically improved accuracies of 0.41 mm and 1.0°, along with a reduced mean targeting time of 56.5 s. Similar results have been published in visualizations and displays outside of the medical/surgical context, suggesting generally that 3D views are more accurate for targeting tasks when combined with 2D visual cues [11,16]. This result may not extend to head-mounted AR displays, which could better portray 3D information than a monitor [17], or perhaps even for displays mounted directly on the surgical instrument [18].

We also found that the alignment of the visualization with respect to the model's physical orientation had a significant impact on navigation time, indicated by the statistically longer total time required for the 'rotated bullseye' view (mean: 60.4 s) compared to the bullseye (mean: 29.3 s). Interestingly the time to threshold (within 3.65 mm and 5°) for the 'rotated bullseye' view was not statistically longer (mean: 22.8 s) than the other 2D visualizations, suggesting that participants could quickly navigate to the general target region but needed a much longer time to make fine adjustments for accurate positioning (Fig. 3). Although perhaps not an unexpected result, it does confirm that navigation systems should ensure that chosen visualizations are aligned with the expected patient orientation to reduce the mental burden for the user and thus achieve more efficient targeting.

With the exception of the 3D visualization, all methods achieved statistically equivalent accuracies suggesting that users were able to accurately navigate (to within 1 mm and 1°) despite the stark differences between visualizations and differences in personal preferences for visualizations. Therefore, we believe that the views designed in this study provided sufficient information for accurate targeting, even if participants required more time to adjust to more challenging views.

Given that there are clear effects of different visualization techniques on task performance times on novice users, we plan to extend this study by recruiting surgeons who have expertise in pelvic fracture fixation surgeries as well as in navigated surgery. This would allow us to more directly conclude which visualizations are better suited for surgeons using surgical navigation, even though we do not expect a difference in performance based on expertise given that targeting with real-time visual feedback significantly reduces the need for 'expert-level knowledge' of the task. Furthermore, while this study focused on targeting, a real surgical procedure would also involve other tasks such as planning a screw position, assessing the planned screw position intraoperatively, and drilling into tissue and bone, all of which may require different visual cues and contextual information for success. For instance, planning a screw target requires much more information from the CT scan than simple targeting. Therefore, we will also extend this study by asking expert surgeons to plan screw targets using navigation and the visualizations proposed here, to determine which different visualizations best augment a surgeon's understanding of the patient's anatomy. Additionally, as AR devices become more portable, accessible and suitable for the OR, we plan to investigate if using similar visualizations on an AR device improves navigation performance.

Overall, we found that the presence of a bullseye view (either on its own or in combination with other visualizations) allows for much more efficient navigation, and can help improve targeting accuracy for novice users. These results suggest that surgical navigation systems implement such a visualization, potentially in tandem with a 'target-fixed' view which could add value during planning or drilling tasks when contextual information from the patient's CT is required to more safely execute the task.

References

1. Carbone, M., Piazza, R., Condino, S.: Commercially available head-mounted displays are unsuitable for augmented reality surgical guidance: a call for focused research for surgical applications (2020). http://journals.sagepub.com/doi/10.1177/1553350620903197
2. El-Hariri, H., Pandey, P., Hodgson, A.J., Garbi, R.: Augmented reality visualisation for orthopaedic surgical guidance with pre- and intra-operative multimodal image data fusion. Healthc. Technol. Lett. **5**, 189–193 (2018)
3. Cartucho, J., Shapira, D., Ashrafian, H., Giannarou, S.: Multimodal mixed reality visualisation for intraoperative surgical guidance. Int. J. Comput. Assist. Radiol. Surg. **15**, 819–826 (2020)
4. Tsukada, S., Ogawa, H., Nishino, M., Kurosaka, K., Hirasawa, N.: Augmented reality-based navigation system applied to tibial bone resection in total knee arthroplasty. J. Exp. Orthop. **6**, 44 (2019)
5. Goerres, J., et al.: Planning, guidance, and quality assurance of pelvic screw placement using deformable image registration. Phys. Med. Biol. **62**, 9018–9038 (2017)
6. Keating, J.F., Werier, J., Blachut, P., Broekhuyse, H., Meek, R.N., O'Brien, P.J.: Early fixation of the vertically unstable pelvis: the role of iliosacral screw fixation of the posterior lesion. J. Orthop. Trauma. **13**, 107–13 (1999)
7. Tonetti, J., et al.: Lésions neurologiques des fractures de l'anneau pelvien. Rev. Chir. Orthop. Reparatrice Appar. Mot. **90**, 122–131 (2004)
8. Zwingmann, J., Hauschild, O., Bode, G., Südkamp, N.P., Schmal, H.: Malposition and revision rates of different imaging modalities for percutaneous iliosacral screw fixation following pelvic fractures: a systematic review and meta-analysis. Arch. Orthop. Trauma Surg. **133**, 1257–1265 (2013)
9. Peng, K.T., et al.: Intraoperative computed tomography with integrated navigation in percutaneous iliosacral screwing. Injury. **44**, 203–208 (2013)

10. Gras, F., Marintschev, I., Wilharm, A., Klos, K., Mückley, T., Hofmann, G.O.: 2D-fluoroscopic navigated percutaneous screw fixation of pelvic ring injuries - a case series. BMC Musculoskelet. Disord. **11**, 153 (2010)

11. Tory, M., Kirkpatrick, A.E., Atkins, M.S., Moller, T.: Visualization task performance with 2D, 3D, and combination displays. IEEE Trans. Vis. Comput. Graph. **12**, 2–13 (2006)

12. Mert, A., et al.: Brain tumor surgery with 3-dimensional surface navigation. Neurosurgery (2012)

13. Fedorov, A., et al.: 3D Slicer as an image computing platform for the Quantitative Imaging Network. Magn. Reson. Imaging. **30**, 1323–1341 (2012)

14. Ma, B., Banihaveb, N., Choi, J., Chen, E.C.S., Simpson, A.L.: Is pose-based pivot calibration superior to sphere fitting? In: Webster, R.J. and Fei, B. (eds.) Medical Imaging 2017: Image-Guided Procedures, Robotic Interventions, and Modeling. p. 101351U (2017)

15. Ungi, T., Lasso, A., Fichtinger, G.: Open-source platforms for navigated image-guided interventions. Med. Image Anal. **33**, 181–186 (2016)

16. Smallman, H.S., John, M.S., Oonk, H.M., Cowen, M.B.: Information availability in 2D and 3D displays. IEEE Comput. Graph. Appl. **21**, 51–57 (2001)

17. Sadda, P., Azimi, E., Jallo, G., Doswell, J., Kazanzides, P.: Surgical navigation with a head-mounted tracking system and display. In: Studies in Health Technology and Informatics, pp. 363–369. IOS Press (2013)

18. Herrlich, M., et al.: Instrument-mounted displays for reducing cognitive load during surgical navigation. Int. J. Comput. Assist. Radiol. Surg. **12**, 1599–1605 (2017)

Prediction of Type II Diabetes Onset with Computed Tomography and Electronic Medical Records

Yucheng Tang[1(✉)], Riqiang Gao[1], Ho Hin Lee[1], Quinn Stanton Wells[2],
Ashley Spann[2], James G. Terry[2], John J. Carr[2], Yuankai Huo[1], Shunxing Bao[1],
and Bennett A. Landman[1,2]

[1] Vanderbilt University, Nashville, TN, USA
yucheng.tang@vanderbilt.edu
[2] Vanderbilt University Medical Center, Nashville, TN, USA

Abstract. Type II diabetes mellitus (T2DM) is a significant public health concern with multiple known risk factors (*e.g.*, body mass index (BMI), body fat distribution, glucose levels). Improved prediction or prognosis would enable earlier intervention before possibly irreversible damage has occurred. Meanwhile, abdominal computed tomography (CT) is a relatively common imaging technique. Herein, we explore secondary use of the CT imaging data to refine the risk profile of future diagnosis of T2DM. In this work, we delineate quantitative information and imaging slices of patient history to predict onset T2DM retrieved from ICD-9 codes at least one year in the future. Furthermore, we investigate the role of five different types of electronic medical records (EMR), specifically 1) demographics; 2) pancreas volume; 3) visceral/subcutaneous fat volumes in L2 region of interest; 4) abdominal body fat distribution and 5) glucose lab tests in prediction. Next, we build a deep neural network to predict onset T2DM with pancreas imaging slices. Finally, motivated by multi-modal machine learning, we construct a merged framework to combine CT imaging slices with EMR information to refine the prediction. We empirically demonstrate our proposed joint analysis involving images and EMR leads to 4.25% and 6.93% AUC increase in predicting T2DM compared with only using images or EMR. In this study, we used case-control dataset of 997 subjects with CT scans and contextual EMR scores. To the best of our knowledge, this is the first work to show the ability to prognose T2DM using the patients' contextual and imaging history. We believe this study has promising potential for heterogeneous data analysis and multi-modal medical applications.

Keywords: Type II diabetes · Electronic medical records · Computed tomography · Metabolic syndrome · Disease onset prediction

1 Introduction

Type II diabetes mellitus (T2DM) [1–3] is a common and significant chronic disease with both inherent and environmental causes. T2DM is characterized by obesity with attendant risk factors including hyperglycemia, hypertension, and hyperglycemia stemming from

© Springer Nature Switzerland AG 2020
T. Syeda-Mahmood et al. (Eds.): ML-CDS 2020/CLIP 2020, LNCS 12445, pp. 13–23, 2020.
https://doi.org/10.1007/978-3-030-60946-7_2

insulin resistance [4–6]. Potential markers of T2DM include the aforementioned risk factors as well as regional obesity and pancreas changes which can be learned from patients' imaging and diagnostic history [7]. Clinical framing of these variables relative to T2DM are complex through multiple risk factors, *e.g.*, body mass index (BMI), pancreas tissue volume, visceral/subcutaneous fat distribution, and glucose tests. Previous works have shown these hand-crafted features can be used to classify the presence of T2DM [8, 9]. However, the longer-term effects of risk factors are less well understood.

Clinical evaluation of patients with potential risk is performed by examining their electronic medical records (EMR) including 1) demographics, 2) ICD-9 codes, 3) lab tests, or 4) clinical and medication histories. With the advent of EMR, researchers have used machine learning and data mining methods in diabetes research, such as predictive biomarker identification, disease prediction, and diagnosis [10]. Mani et al. used demographic, clinical and lab parameters from EMR with different machine learning algorithms (linear based, one sample-based, decision tree based, and one kernel-based classifiers) for predicting T2DM risks in the period six months to one-year prior to diagnosis of diabetes [8]. Zheng et al. proposed a framework for identifying subjects with and without T2DM from EMR via feature engineering and similar machine learning methods [11]. Anderson et al. employed logistic regression and a random-forests probabilistic model on patients' full EMR or restricted EMR and showed EMR phenotyping can out-perform predict diagnosis of T2DM with conventional screening methods [7]. Meanwhile, there are few researches applying medical imaging studies (MRI, abdominal CT) to understand the association between T2DM and tissue composition as well as volumetric measurements. For instance, livers volumes, size of pancreas and body fat content [12, 13] are related to T2DM.

Recent advances, such as use of multi-modal machine learning [14, 15], bring opportunities for medical applications derived from both EMR and medical image data [16, 17]. Virostko et al. discovered pancreas volume declines with disease duration in type I diabetes (T1D) patients by using electronic medical record and magnetic resonance imaging (MRI)/ CT scans [9]. Chaganti et al. developed a method for multi-modal big data studies in MRI image processing that used EMR information, to classify diabetes patients in orbit CT [18]. However, the above works do not focus on early prognosis. Since the abdominal CT is becoming a routinely acquired imaging technique [19, 20], our goal is to explore the feasibility of combining EMR and CT to predict risk factors for T2DM a year prior to diagnosis.

In this work, we investigate five different types of EMR features to predict the onset of T2DM. We extract the 1) patient demographics, 2) pancreas volume, 3) fat volume in L2 region of interests (ROI), 4) visceral/subcutaneous fat distribution along the abdomen, and 5) glucose lab tests from each patient's clinical history. We perform each configuration of EMR in an ablation scheme for T2DM onset prediction. For each subject, we formulate an EMR feature vector describing their clinical history from each configuration. We show that each contextual feature from patients' clinical history improves the prediction of onset T2DM. Next, we construct a deep neural network for encoding pancreas CT imaging slices for the T2DM onset prediction. Inspired by previous works on EMR-guided image processing [8], we developed a framework that combines CT images and EMR features.

We conduct experiments on 997 subjects using the "case-control" design [21, 22]. 401 subjects are with diagnoses of T2DM. The remaining 596 subjects are in the control group of non-diabetes. For extracting anatomic volumes, we trained the segmentation model for this work using 3D U-net with 100 CT scans and labels from multi-atlas labeling MICCAI challenge 2015 [23]. We believe this work could motivate further investigation of heterogenous data analysis and EMR guided multi-modal medical applications.

In summary, our contributions in this work are: 1) We present a "case-control" study design for onset prediction of T2DM; 2) We evaluate that five different configurations of EMR features can contribute to prediction of onset T2DM a year prior to diagnosis; and 3) We show that EMR-image multi-modal framework for heterogenous data analysis improves predictive power for medical applications.

2   Method

Our proposed method involves five different categories of EMR features and a deep neural network architecture for encoding pancreas imaging slices, as illustrated in Fig. 1.

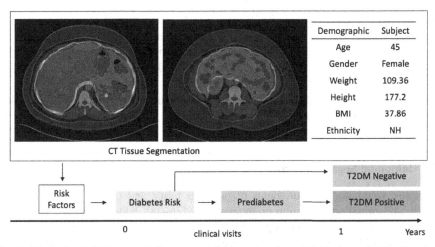

Fig. 1. Example of CT images (subcutaneous fat in navy, visceral fat in brown). The flowchart shows from risk factors to T2DM diagnoses with multiple modalities (CT and demographic) at least a year ahead of diagnoses. (Color figure online)

2.1 Data of T2DM Studies

A total of 997 de-identified subjects were selected and retrieved from our medical center under the institutional review board (IRB) approval from 6317 studies with diagnosis codes involving spleen abnormalities (cohort A). non-spleen abnormalities (cohort B). The dataset follows "case-control" ICD-9 code design [22]: (1). Case: T2DM subjects identified from ICD-9 code with diagnostic date (*ICD-9 = 250.## group of type 2 diabetes diagnosis*); (2) Control: non-diabetes subjects without diabetes ICD-9 codes or medication. There are 401 cases of T2DM diagnosis and 596 subjects under the control group of non-diabetes who were chosen for having similar imaging availability. Meanwhile, all CT scans are with contrast enhancement in portal venous phase. The in-plane pixel dimension of each CT scans varies from 0.7 to 1.2 mm. Each image is preprocessed by excluding outlier intensities beyond −1000 and 1000 HU. Images slice thickness ranges from 2 to 4 mm. Each CT scans consists of 60 to 200 slices of 512 × 512 pixels. For consistency consideration, we pre-processed all CT scans with a soft tissue window with a range of [−175, 250] HU, the window effect is studied in [24]. Intensities were normalized to [0, 1].

Pancreas Imaging Slices: Clinically acquired CT scans usually have large variance in field of view. We implemented a pre-processing step with body part regression, which is a method to illustrate abdominal intrinsic structure [25]. The body part regression assists to automatically remove slices based on inconsistent volumes, and to localize abdominal anatomies. We adopted the pre-trained model from the unsupervised regression network [25] to find slices in the pancreas region (scalar reference index score ranges from −1 to 1 as indicated in [25]).

Timeline of T2DM diagnosis and CT sessions: To clarify the task of predicting future risk of T2DM, we obtained longitudinal CT sessions along with ICD-9 T2DM diagnosis code and date. We first assured the diagnosis date for each T2DM patient, then retrieved a CT session and EMR at least one year ahead of the ICD-9 date. A randomized controlled trial for a CT session and EMR was collected from a non-T2DM control group. The time interval between diagnosis date and CT session date ranges from 365 to 1690 days (mean: 456, median: 540). Only one CT session per patient is used in the study.

2.2 Abdominal Segmentations

To acquire volumetric measurements of pancreas tissue and obesity of patients, we perform abdominal segmentation on the pancreas [26–28], body wall/mask, and body fat from CT imaging [29]. In this paradigm, we use a dataset of 100 subjects from the MICCAI 2015 Multi-Atlas Labeling Challenge with 12 anatomies labels annotated by experts. In detail, we train a 3D multi-organ segmentation network [30] for segmenting the pancreas and inner/outer abdomen wall by multi-task learning. Each scan is down-sampled from [512, 512] to [168, 168] and normalized to consistent voxel resolution of [2 × 2 × 6]. The output and ground truth labels are compared using Dice loss [31]. We ignored the background loss in order to increase weights for anatomies. The final segmentation maps are up-sampled to original space with the nearest interpolation in order to spatially align with CT resolution. This approach is trained end-to-end, and the resulting segmentation is shown in Fig. 2. In order to extract fat volume measurements, we perform fuzzy c-means [32] on CT images for body mask and fat segmentation in an unsupervised scheme.

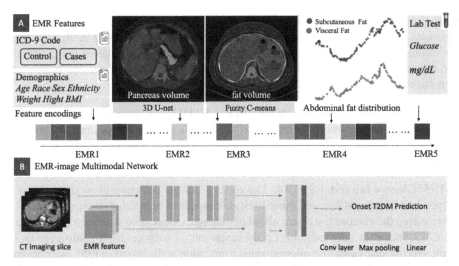

Fig. 2. Method framework. Part A indicates EMR features of five configurations. ICD-9 code is used as exclusion criteria for case-control design. The demographics is used as the base feature vector (EMR1). Pancreas and fat segmentation imply the volumetric measurements for EMR2 and EMR3, respectively. The abdominal fat distribution and glucose lab test are used to compose the EMR4 and EMR5 features. Part B shows our network architecture for combining EMR features and CT imaging slices for prediction of onset T2DM.

2.3 EMR Feature Extraction

A clinical EMR captured patient demographics, clinical notes, lab tests, medication codes, diagnosis codes, treatment procedures. For our cohort, we extracted the de-identified clinical phenotype base on a hierarchical categorization of ICD-9 (International Classification of Disease - 9) codes [33]. A phecode-groups a set of ICD-9 codes to imply a diagnosis category. For example, type II diabetes mellitus includes *250.00, 250.02, 250.40, 250.42, ..., etc*. We use an open-sourced PyPhewas tool to automatically generate the ICD-9 categories [17], given a list of clinical visits.

In this study, we investigate five different configurations of EMR features in predicting onset T2DM in an ablation study scheme:

EMR1: Demographic. We extract demographic history for each de-identified patient involves age, sex, race, ethnicity, weight, height, and BMI. A vector of length 7 is formed for each subject. The risk factors are normalized from 0 to 1.

EMR2: Pancreas volume + EMR1. Pancreatic necrosis and acute pancreatitis mortality are major factors of insulin insufficiency. Herein, we calculate pancreas volume for each patient from the CT session for one year prior to diagnoses at predicting T2DM. The volumetric measurements are acquired from pancreas segmentation using 3D U-net as shown in Sect. 2.2.

EMR3: Fat volume in L2 region + EMR2. Obesity increases the risk of systemic inflammatory response syndrome on T2DM prediction. Abdominal fat represented by waist circumference is better than single weight or BMI at characterizing metabolic syndrome [34]. We extract visceral and subcutaneous fat volume along the L2 section. Fat volumes are acquired from automatic fat/abdomen wall segmentation and distribution regressed by body part regression [25], where L2 slices are retrieved by body part regression scores in range -1 to 0.

EMR4: Abdominal fat volume distribution + EMR3. To identify the utility of abdominal fat in prediction, we evaluated abdominal visceral fat volume and subcutaneous fat volume. Similar to EMR3, we extract abdominal slices by BPR scores in range of -6 to 5.

EMR5: Glucose lab test + EMR4. Lab testing enables detection of prediabetes and suggestions for weight loss and other lifestyle changes to help delay or prevent T2DM. We retrieved the glucose test a year ahead of diagnoses associated with each patient into the EMR vector. In our study, 17 T2DM cases and 87 subjects from control are lack of the glucose test value, the missing value imputation is discussed in [35]. We present mean imputation that simply calculate the mean of the observed values for individuals who are non-missing from T2DM cases and controls, respectively.

In order to project unimodal representation using the heterogeneity of multimodal data, we use a joint representation learning framework. In this work, we address the joint training problem by introducing an EMR-image multimodal network. As shown in Fig. 2, we include neural network for each modality that fused before the last fully connected and activation layer. Next, we fuse them at the last hidden layer, where the posteriors obtain the correlation between projected non-linear spaces of two modalities. Then this feature space is fixed and passed through a shallow linear layer to the targets. The networks are optimized by cross-entropy objective and stochastic gradient descent. The network comprises four 3D conv layers, followed by ReLU and pooling layers. The whole framework is trained in end-to-end manner and results in multimodal fusion directly.

3   Experiments

3.1   Experimental Design

To perform each EMR configuration as well as bilinear multimodal network, we implemented experiments with cross validation on 997 subjects from the "case-control" study. To perform standard five-fold cross validation, we split dataset into five complementary folds, each of which contains 200 cases (197 in last fold). For each fold evaluation, we use three folds as training, one fold as validation, and remaining as testing.

3.2   Implementation Details and Metric

The EMR-image multimodal network consists of four levels of convolution blocks, each block has a $3 \times 3 \times 3$ convolutional layer, followed by rectified linear units (ReLU) and a max-pooling of $2 \times 2 \times 2$ and strides of 2. The learning rate for the EMR-image multimodal network is set with $1e-5$. We use a batch size of 1 for all implementations and adopted the ADAM algorithm with SGD, momentum = 0.9. Implementations are performed using Pytorch 1.0 with NVIDIA Titan X GPU 12G memory and CUDA 9.0. The code of all experiments, including baseline methods, is implemented in Python 3.6 with Anaconda3. The area under the curve (AUC) is calculated from the receiver operating curve (ROC) for each experiment. ROC curves are generated by the true positive rate (TPR) against the false positive rate (FPR) at various threshold settings. AUC measures the entire two-dimensional area underneath the entire ROC curve.

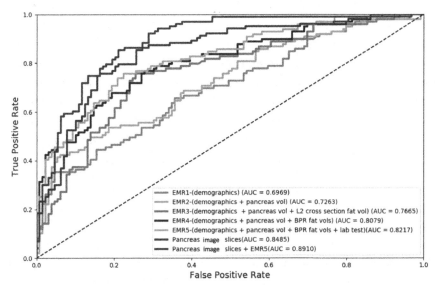

Fig. 3. The receiver operating characteristic (ROC) curves of results on the prediction of onset T2DM. Area Under Curve (AUC) is shown in right bottom of figure. EMR features from 1 to 5 achieve barely satisfactory results, the use of pancreas slice and the proposed multimodal network performs better.

3.3 Results and Analyses

The experimental result of all settings is shown in Table 1. We summarize the mean and standard deviation in accuracy, AUC, F1, recall, and precision for comparison. In the context of five EMR configurations, we can see that each risk factor contributes to improvements. In particular, improvements over demographics increases as risk factors get more diversity from images. We observe a higher improvement in adding body fat distribution. We also see consistent better performance using pancreas slices, showing the advantage of imaging biomarkers. In the multimodal analysis, we achieve the best accuracy at 86.19%, combining the pancreas slice and EMR5. Comparing with EMR5 or pancreas slices, our method achieved consistently better result, with 6.93% and 4.25% in AUC respectively. These imply the effectiveness of using heterogeneous data.

Figure 3 shows the ROC-AUC curves for all experiments. The green curve indicates the combination of EMR and pancreas slices in the training, we show that EMR-image multi-modal framework for heterogenous data analysis improves predictive power (AUC 0.8910) compare to brown curve (AUC 0.8485) and yellow curve (AUC 0.8217).

Table 1. Result on the mean and standard deviation (std) (%, average (std) of five-fold cross validation). The *p-values* are calculated using McNemar test with respect to AUC in the row above each test. * indicates significant with $p < 0.01$.

Method	Accuracy	AUC	F1	Recall	Precision
EMR1	67.25(2.97)	69.69(2.45)	51.93(2.99)	39.28(4.64)	77.98(4.34)
EMR2	70.04(2.58)	72.63(2.13)	56.74(3.08)	49.72(4.04)	79.01(3.98)
EMR3	74.21(2.01)	76.65(1.75)*	59.82(2.76)	54.85(4.02)	80.06(3.34)
EMR4	76.93(1.96)	80.79(1.42)*	64.59(2.51)	59.81(3.54)	80.96(3.51)
EMR5	80.02(1.98)	82.17(1.31)*	67.42(2.43)	61.73(2.98)	82.79(3.64)
Pan slice	82.46(1.70)	84.85(1.19)	72.54(1.89)	64.49(2.74)	84.37(3.72)
Pan slice + EMR5	**86.19(1.45)**	**89.10(1.10)***	**76.85(1.72)**	**67.61(1.98)**	**87.62(3.32)**

*The best average results are shown in **bold**.* Pan slice indicates pancreas slice.

4   Discussion and Conclusion

In this work, we targeted at understanding of the association among T2DM, patient clinical history, tissue composition imaging as well as volumetric measurements. In this paper, we used a "case-control" study design for the onset of T2DM prediction task. We show that direct prediction using either EMR features or CT imaging enables prediction of risk factors for T2DM. We investigate five different configurations of EMR including demographics, pancreas volume measurements, visceral/subcutaneous fat volume, body fat distribution and glucose lab tests. Different factors contribute to improve AUC of the prediction. Meanwhile, we institute a holistic method for EMR features and CT images, and show improved performance over the base EMR features and imaging methods alone. The proposed method builds upon the connections between different modalities (EMR and CT) through which we extrapolate a joint representation learning in the multimodal machine learning setting. The way we obtain the joint feature space represents a general means of exploitation based on multiple modalities. Specifically, jointly using imaging techniques, notes and quantitative measurements offers innovative understandings of the challenge. It also allows us to address problems systematically. We are excited to explore more mechanisms such as bilinear multimodal architecture [36] or dual attention [37], and in particular study the EMR-guided approaches in the future.

One of the limitations in this work focuses on the retrospective EMR study design. The ability of case-control design has been widely evaluated [21]. It sometimes is unable to detect very small relative risks from exposures and it tends to be more effective in large-scale, collaborative, multi-center studies. For example, compared to a previous study with AUC of 0.877 [8] using their institutional data, we achieved slightly higher

AUC in our cohort (Table 1). The absolute performance on different institutional cohorts shows some relative risk factors could be typically regarded as low importance, but these factors may have higher importance in the population at large. Herein, the full cohort review is needed in the future. And to investigate subsequent studies to guarantee differences among controls do not confound results.

Acknowledgements. This research is supported by Vanderbilt-12Sigma Research Grant, NSF CAREER 1452485, NIH 1R01EB017230 (Landman). This study was in part using the resources of the Advanced Computing Center for Research and Education (ACCRE) at Vanderbilt University, Nashville, TN. The imaging dataset(s) used for the analysis described were obtained from ImageVU, a research resource supported by the VICTR CTSA award (ULTR000445 from NCATS/NIH).

References

1. Hales, C.N., Barker, D.J.P.: Type 2 (non-insulin-dependent) diabetes mellitus: the thrifty phenotype hypothesis. Diabetologia **35**(7), 595–601 (1992). https://doi.org/10.1007/BF0040 0248
2. Chen, L., Magliano, D.J., Zimmet, P.Z.: The worldwide epidemiology of type 2 diabetes mellitus—present and future perspectives. Nat. Rev. Endocrinol. **8**(4), 228–236 (2012)
3. Neeland, I.J., et al.: Dysfunctional adiposity and the risk of prediabetes and type 2 diabetes in obese adults. JAMA **308**(11), 1150–1159 (2012)
4. Tognini, G., Ferrozzi, F., Bova, D., Bini, P., Zompatori, M.: Diabetes mellitus: CT findings of unusual complications related to the disease: a pictorial essay. Clin. Imaging **27**(5), 325–329 (2003)
5. Association, A.D.: Diagnosis and classification of diabetes mellitus. Diabetes Care **37**(Supplement 1), S81–S90 (2014)
6. Fletcher, B., Gulanick, M., Lamendola, C.: Risk factors for type 2 diabetes mellitus. J. Cardiovasc. Nurs. **16**(2), 17–23 (2002)
7. Anderson, A.E., Kerr, W.T., Thames, A., Li, T., Xiao, J., Cohen, M.S.: Electronic health record phenotyping improves detection and screening of type 2 diabetes in the general United States population: a cross-sectional, unselected, retrospective study. J. Biomed. Inform. **60**, 162–168 (2016)
8. Mani, S., Chen, Y., Elasy, T.. Clayton, W., Denny, J.: Type 2 diabetes risk forecasting from EMR data using machine learning. In: AMIA Annual Symposium Proceedings, vol. 2012, p. 606. American Medical Informatics Association (2012)
9. Virostko, J., Hilmes, M., Eitel, K., Moore, D.J., Powers, A.C.: Use of the electronic medical record to assess pancreas size in type 1 diabetes. PLoS ONE, **11**(7), e0158825 (2016)
10. Kavakiotis, I., et al.: Machine learning and data mining methods in diabetes research. Comput. Struct. Biotechnol. J. **15**, 104–116 (2017)
11. Zheng, T., et al.: A machine learning-based framework to identify type 2 diabetes through electronic health records. Int. J. Med. Informatics **97**, 120–127 (2017)
12. Garcia, T.S., Rech, T.H., Leitao, C.B.: Pancreatic size and fat content in diabetes: a systematic review and meta-analysis of imaging studies. PLoS ONE **12**(7), e0180911 (2017)
13. Vu, K.N., Gilbert, G., Chalut, M., Chagnon, M., Chartrand, G., Tang, A.: MRI-determined liver proton density fat fraction, with MRS validation: comparison of regions of interest sampling methods in patients with type 2 diabetes. J. Magn. Reson. Imaging **43**(5), 1090–1099 (2016)

14. Zhang, Z., Chen, P., Shi, X., Yang, L.: Text-guided neural network training for image recognition in natural scenes and medicine. IEEE Trans. Pattern Anal. Mach. Intell. (2019)
15. Baltrušaitis, T., Ahuja, C., Morency, L.-P.: Multimodal machine learning: a survey and taxonomy. IEEE Trans. Pattern Anal. Mach. Intell. **41**, 2 (2018)
16. Evans, J.A.: Electronic medical records system. ed: Google Patents (1999)
17. Chaganti, S., Bermudez, C., Mawn, L.A., Lasko, T., Landman, B.A.: Contextual deep regression network for volume estimation in orbital CT. In: Shen, D., Liu, T., Peters, T.M., Staib, L.H., Essert, C., Zhou, S., Yap, P.-T., Khan, A. (eds.) MICCAI 2019. LNCS, vol. 11769, pp. 104–111. Springer, Cham (2019). https://doi.org/10.1007/978-3-030-32226-7_12
18. Chaganti, S., et al.: Electronic medical record context signatures improve diagnostic classification using medical image computing. IEEE J. Biomed. Health Inform. **23**(5), 2052–2062 (2018)
19. Tang, Y., et al.: Contrast phase classification with a generative adversarial network. arXiv preprint arXiv:1911.06395 (2019)
20. Kulama, E.: Scanning protocols for multislice CT scanners. Br. J. Radiol. **77**(suppl_1), S2–S9 (2004)
21. Crombie, I.K.: The limitations of case-control studies in the detection of environmental carcinogens. J. Epidemiol. Community Health **35**(4), 281–287 (1981)
22. Mann, C.: Observational research methods. Research design II: cohort, cross sectional, and case-control studies. Emer. Med. J. **20**(1), 54–60 (2003)
23. Landman, B., Xu, Z., Igelsias, J., Styner, M., Langerak, T., Klein, A.: MICCAI Multi-Atlas Labeling Beyond the Cranial Vault–Workshop and Challenge (2015)
24. Huo, Y., et al.: Stochastic tissue window normalization of deep learning on computed tomography. J. Med. Imaging **6**(4), 044005 (2019)
25. Yan, K., Lu, L., Summers, R.M.: Unsupervised body part regression via spatially self-ordering convolutional neural networks. In: 2018 IEEE 15th International Symposium on Biomedical Imaging (ISBI 2018), pp. 1022–1025. IEEE (2018)
26. Xu, Z., et al.: Efficient multi-atlas abdominal segmentation on clinically acquired CT with SIMPLE context learning. Med. Image Anal. **24**(1), 18–27 (2015)
27. Xu, Y., et al.: Outlier Guided Optimization of Abdominal Segmentation. arXiv2002.04098
28. Wang, Y., Zhou, Y., Shen, W., Park, S., Fishman, E.K., Yuille, A.L.: Abdominal multi-organ segmentation with organ-attention networks and statistical fusion. Med. Image Anal. **55**, 88–102 (2019)
29. Xu, Z., Baucom, R.B., Abramson, R.G., Poulose, B.K., Landman, B.A.: Whole abdominal wall segmentation using augmented active shape models (AASM) with multi-atlas label fusion and level set," in Medical Imaging 2016: Image Processing, 2016, vol. 9784: International Society for Optics and Photonics, p. 97840U
30. Çiçek, Ö., Abdulkadir, A., Lienkamp, Soeren S., Brox, T., Ronneberger, O.: 3D U-Net: learning dense volumetric segmentation from sparse annotation. In: Ourselin, S., Joskowicz, L., Sabuncu, Mert R., Unal, G., Wells, W. (eds.) MICCAI 2016. LNCS, vol. 9901, pp. 424–432. Springer, Cham (2016). https://doi.org/10.1007/978-3-319-46723-8_49
31. Sudre, Carole H., Li, W., Vercauteren, T., Ourselin, S., Jorge Cardoso, M.: Generalised dice overlap as a deep learning loss function for highly unbalanced segmentations. In: Cardoso, M.Jorge, et al. (eds.) DLMIA/ML-CDS -2017. LNCS, vol. 10553, pp. 240–248. Springer, Cham (2017). https://doi.org/10.1007/978-3-319-67558-9_28
32. Bezdek, J.C., Ehrlich, R., Full, W.: FCM: the fuzzy c-means clustering algorithm. Comput. Geosci. **10**(2–3), 191–203 (1984)
33. Quan, H., et al.: Coding algorithms for defining comorbidities in ICD-9-CM and ICD-10 administrative data. Med. Care **43**, 1130–1139 (2005)
34. Carey, V.J., et al.: Body fat distribution and risk of non-insulin-dependent diabetes mellitus in women: the Nurses' Health Study. Am. J. Epidemiol. **145**, 7 (1997)

35. Baraldi, A.N., Enders, C.K.: An introduction to modern missing data analyses. J. Sch. Psychol. **48**(1), 5–37 (2010)
36. Mroueh, Y., Marcheret, E., Goel, V.: Deep multimodal learning for audio-visual speech recognition. In: 2015 IEEE International Conference on Acoustics, Speech and Signal Processing (ICASSP), pp. 2130–2134. IEEE (2015)
37. Zhang, Z. , Chen, P., Sapkota, M., Yang, L.: Tandemnet: Distilling knowledge from medical images using diagnostic reports as optional semantic references. In International Conference on Medical Image Computing and Computer-Assisted Intervention (2017)

A Radiomics-Based Machine Learning Approach to Assess Collateral Circulation in Ischemic Stroke on Non-contrast Computed Tomography

Mumu Aktar[1]($\boxtimes$) , Yiming Xiao[1] , Donatella Tampieri[2], Hassan Rivaz[3] ,
and Marta Kersten-Oertel[1]

[1] Department of Computer Science and Software Engineering, Concordia University,
1455 boul. De Maisonneuve O., Montreal, QC H3G 1M8, Canada
m_ktar@encs.concordia.ca
[2] Department of Diagnostic Radiology, Kingston Health Sciences Centre,
Kingston General Hospital, Kingston, ON K7L 2V7, Canada
[3] Department of Electrical and Computer Engineering, Concordia University,
1455 boul. De Maisonneuve O., Montreal, QC H3G 1M8, Canada

Abstract. Assessment of collateral circulation in ischemic stroke, which
can identify patients for the most appropriate treatment strategies, is
currently conducted with visual inspection by a radiologist. Yet numer-
ous studies have shown that visual inspection suffers from inter and
intra-rater variability. We present an automatic evaluation of collaterals
using radiomic features and machine learning based on the ASPECTS
scoring terminology with non-contrast computed tomography (NCCT).
The method includes ASPECTS regions identification, extraction of
radiomic features, and classification of collateral scores with support
vector machines (SVM). Experiments are performed on a dataset of 64
ischemic stroke patients to classify collateral circulation as good, intermediate, or poor and yield an overall area under the curve (AUC) of 0.86
with an average sensitivity of 80.33% and specificity of 79.33%. Thus, we
show the feasibility of using automatic evaluation of collateral circulation using NCCT when compared to the ASPECTS score by radiologists
using 4D CT angiography as ground truth.

Keywords: Ischemic stroke · Collateral circulation · Computer-aided
diagnosis · Non-contrast CT · Machine learning · Radiomics

1 Introduction

Acute ischemic stroke (AIS) caused by blocked arteries in the brain is one of
the leading causes of death worldwide. Treatment strategies of AIS vary from
intravenous tissue plasminogen activator (IV-tPA) to endovascular thrombec-
tomy treatments (EVT) based on the time window and patients' conditions.

© Springer Nature Switzerland AG 2020
T. Syeda-Mahmood et al. (Eds.): ML-CDS 2020/CLIP 2020, LNCS 12445, pp. 24–33, 2020.
https://doi.org/10.1007/978-3-030-60946-7_3

EVT is one of the best treatments for restoring blood flow through blocked arteries but its success rate depends on the extents of a patient's collateral circulation. Collateral circulation is a secondary vascular network that is recruited temporarily that allows for the survival of viable brain tissues when the main conduits fail due to ischemic stroke. The extent of leptomeningeal collateral flow from the middle cerebral artery (MCA) flowing to the anterior cerebral artery (ACA) and posterior cerebral artery (PCA) has been shown to be a radiologic surrogate predicting the response of revascularization therapy [12]. However, scoring of the collaterals manually following conventional radiologic strategy suffers from the intra- and inter-rater variability [2,9,10], less reliable results, and is time-consuming. Some studies have compared automated approach with visual inspection based on ASPECTS evaluation having greater agreement (κ = 0.90) than neuroradiologists [18] or slightly worse agreement than human expert ratings [14]. Therefore, developing automatic and robust approaches to collateral evaluation in AIS based on systematic radiologic criteria and methods is important. A number of different approaches have been proposed to score collaterals, for example, ASPECTS (Alberta Stroke Program Early CT Score) [24], the collateral score of the Society of NeuroInterventional Surgery/Society of Interventional Radiology (ASITN/SIR) based on conventional angiography [11] which is adapted to be applicable to dynamic computed tomography angiography (CTA) further in the study of Sekar et al. [26], the scores of Christoforidis et al. [6], Miteff System [20], Mass System [17], modified Tan Scale [29], regional leptomeningeal collateral score (rLMC) [19], collateral evaluation with 4D-CTA based on ASPECTS [15,31], and ACCESS [2].

ASPECTS is one of the most reliable, systematic and robust approaches shown to have positive clinical outcomes in ischemic damage detection in many studies (e.g. [7,16,25–28]) with baseline CTA source images(CTA-SI), CT perfusion images (CTP), contrast-enhanced CT (CECT), non-contrast CT (NCCT), and timing invariant CTA (TiCTA) modalities respectively. Although some studies [3,5,25] show that contrast-enhanced CT has superior performance delineating brain vasculature, these are limited to manual intervention or semi-quantitative approaches. Different studies (Brainomix: e-ASPECTS, e-CTA, iSchemaView: Rapid CTA, Rapid ASPECTS, Syngo.via Frontier ASPECT Score Prototype V2: not clinically approved) have focused on automating ASPECTS using artificial intelligence and feature-based machine learning. The e-ASPECTS software from Brainomix Ltd. (Oxford, UK) and RAPID ASPECTS by iSchemaView (Menlo Park, USA) are the only two certified clinical software for ischemic damage detection using ASPECTS on NCCT. Although they are not intended yet to be used as stand-alone diagnostic tools, both suggested NCCT as an alternative to CTP for ischemic damage quantification [21]. The ability of NCCT to work as the stand-alone diagnostic tool extracting much clinical information was shown by Sheih et al. [27]. In their work, the authors compared diffuse hypoattenuation and focal hypoattenuation on contralateral hemispheres in 10-ASPECTS regions and obtained an area under the curve (AUC) of 90.2% for a total of 103 subjects. The study by Kuang et al. [16]

also performed contralateral analysis using a machine learning-based approach to assess early ischemic changes by classifying the 10-ASPECTS regions based on the differences of contralateral texture features. Taking diffusion-weighted imaging (DWI) as ground truth, this study with 257 patients obtained an AUC of 0.89 between the proposed method and experts' reading. Our work resembles Kuang *et al.* [16] in that it uses contralateral radiomic features, however unlike their work we evaluate collateral circulation rather than early ischemic damage.

Since NCCT is easily available and used as a front-line diagnostic tool in clinical settings, also being free from contrast agent that can cause adverse effects to some patients, we used NCCT to automatically assess scoring in acute ischemic stroke based on the ASPECTS terminology. Unlike most of the state-of-the-art methods of automating ASPECTS to obtain ischemic damage by assessing hypoattenuation using DTI as ground truth, we used 4D CTA as ground truth to evaluate collaterals scored through observing multiple phases by radiologists. We aim to evaluate collaterals using NCCT with radiomic feature extraction in the MCA territory of left/right hemispheres and classify them into good, intermediate, or poor categories with support vector machines (SVM). Since brain collaterals vary between individuals and represent symmetric characteristics between left/right hemispheres of the same individual, we extracted radiomic features from each side of the hemispheres of a subject separately and took the difference between them to obtain the non-symmetry. Figure 1 shows the different collateral categories in contrast-enhanced CTA. To the best of our knowledge, this is the first study using NCCT to evaluate the collateral circulation. The underlying assumption of our novel approach is that we can identify regions with insufficient collateral circulation using radiomic features based on tissue degeneration which may be captured on NCCT and score the extent of collaterals using these features.

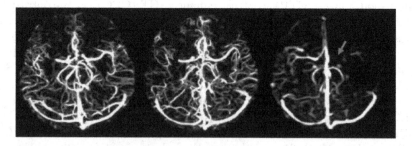

Fig. 1. From left to right, an example of good, intermediate and poor collaterals on contrast-enhanced CTA. The blue arrow indicates the occlusion on the MCA.

2   Materials and Methods

2.1   Scanning Protocols

We have evaluated our method with 8 poor, 17 intermediate, and 39 patients with good collaterals. All 64 subjects underwent imaging at the Montreal Neurological Hospital (Montreal, Canada). A Toshiba Aquilion ONE 320-row detector 640-slice cone-beam CT (Toshiba medical systems, Tokyo, Japan) scanner, which provides whole-brain perfusion and dynamic vasculature information in one single examination with a single rotation of the gantry, was used to capture the 4D CTA. A series of intermittent volume scans are performed for 60 s with a scanning speed of 0.75 s/rotation to capture a total of 19 volumes for each patient with low-dose scanning for every 2 s during the arterial phase and 5 s during the venous phase. The 19 volumes are divided into 5 groups according to the tube current. The first volume, which we used in our method, is captured before contrast arrival. Isovue-370 (Iopamidol) was used as the non-ionic and low osmolar contrast medium to visualize the vessels in the remaining volumes.

2.2 Ground Truth Labels

The ground truth for collateral circulation was based on scoring by two radiologists examining the 4D CTA images. Following the ASPECTS scoring terminology, the radiologists' defined patients with 0–50% collaterals as poor, greater than 50% and less than 100% as intermediate and 100% collaterals as good.

2.3 Mapping of ASPECTS Regions

Prior to the collateral evaluation, we employed atlas registration to align the 10-ASPECTS regions in all patients. The atlas is generated by extracting the ASPECTS regions from the MNI structural atlas and Harvard Oxford atlas using FSL. Following the ASPECTS score in acute stroke [1] and the study of Pexman et al. [24], we extracted the insular ribbon (I), caudate (C), Lentiform nucleus (L), internal capsule (IC), M1 (anterior to the anterior end of the Sylvian fissure including frontal operculum), M2 (anterior temporal lobe), M3 (posterior temporal lobe), M4 (Anterior MCA territory), M5 (Lateral MCA territory) and M6 (Posterior MCA territory). These are rostral to basal ganglia and approximately 2cm superior to M1, M2 and M3 respectively.

Three steps were performed to map the atlas to the subject brain. First, as the atlas we extracted is in MNI template space, it was mapped onto an average CT template (created using 12 healthy subjects' brains following the unbiased group-wise registration approach by Fonov et al. [8]) using affine registration. Next, the CT template was registered to the subjects' native space using symmetric normalization (SyN). Finally, the transformations obtained from the previous steps were used to map the atlas to each subject to delineate specific ASPECTS regions in all subjects. All registration steps were done using Advanced Normalization Tools (ANTs) (stnava.github.io/ANTs). We extracted the brain from each subject using FSL following the study of Muschelli et al. [22]. Figure 2 shows the ASPECTS regions mapped onto a subject.

2.4 Pre-processing

The skull was removed from the NCCT images following the study of Muschelli et al. [22]. In this work each image is thresholded within the brain tissue range

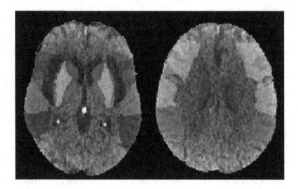

Fig. 2. The 10-ASPECTS regions mapped to an individual patient's brain

of 0–100 Hounsfield units (HU) before skull stripping thus removing any calcifications with very high-intensity values, which were present in some of our patients. Further, a Gaussian pyramid from the multi-scale image representation approach using the kernel from the study of Burt *et al.* [4] was applied to perform smoothing and sub-sampling by one level to all subjects.

2.5   Image Features

The study of Shieh *et al.* [27] shows that ischemic damage can be identified through focal and diffuse hypoattenuation occurring in any ASPECTS region. Following the study of Shieh *et al.*, a deviation and a contrast map were extracted from each side of the brain in order to highlight the areas with insufficient collaterals due to ischemic damage. The deviation and contrast degradation in the areas with less collaterals can be obtained comparing the difference between affected and unaffected sides of the brain. Therefore, the deviation map, D_{map} is generated from each side of the brain by subtracting each voxel's intensity, $V(x, y, z)$ from the mean voxel intensity, V_μ and normalizing it with the standard deviation, V_σ using the Eq. 1.

$$D_{map} = \frac{V(x, y, z) - V_\mu}{V_\sigma} \tag{1}$$

On the other hand, a contrast map is obtained from each side choosing the edges with maximum gradient using the Sobel operator. Rather than comparing the deviation and contrast map between both sides using threshold as the study of Shieh [27], radiomic features are extracted automatically from the maps separately to obtain the spatial relation of voxels and finally the difference between features from each hemisphere is taken (Fig. 3 shows the feature maps with the radiomic features). The feature classes used include radiomic features for the 3D subjects from the study of Van *et al.* [30], as described below.

Gray Level Co-Occurrence Matrix (GLCM): The co-occurrence of voxels for specific values are used to examine the textures of the images by statistical measurement of energy, contrast, entropy, homogeneity, correlation, variance, sum average, autocorrelation, and dissimilarity.

Gray Level Size Zone Matrix (GLSZM): The gray level zone of each subject is quantified by computing the number of voxels that share the same gray level intensity in a 26-connected region.

Global Features: Along with GLCM and GLSZM, the global features of mean, energy, entropy, variance, skewness, and kurtosis of the entire MCA territory from both sides were considered.

In total, 56 features are extracted from the deviation and contrast maps from each side of the brain.

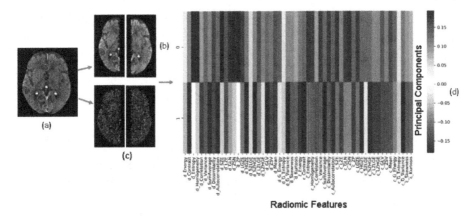

Fig. 3. Feature maps with occlusion and radiomic features (a) Original brain image (b) Deviation maps of left and right hemispheres with highlighted occlusion (c) Contrast maps of both sides with the degradation shown in a polygon (d) Radiomic features (Color figure online)

2.6   Classification of Collaterals

Before feeding the radiomic features obtained from the difference of the hemispheres into a classifier, they were ranked using Principal Component Analysis (PCA) with 97% variance. Figure 3(d) shows the radiomic features' correlation with the first two principal components where the color bar represents the weight of correlation. For example, the GLCM features: *d_Energy, d_Homogeneity* from the first principal component have the highest correlation of 0.19 with the principal component which is also visible from the color bar range. The feature names are distinguished starting with 'd' for deviation map and 'c' for contrast map. These are then fed into the One-Vs-Rest SVM classifier with the radial basis function (RBF) kernel using balanced class weight (to penalize the majority class) to classify collaterals into good, intermediate, or poor cases.

3   Results

We applied k-fold (k = 10) cross-validation and the performance of the developed method was evaluated using receiver operating characteristic (ROC). We obtained an overall area under the curve (AUC) of 0.86 for best sensitivity and specificity with AUC of 0.85 for poor, 0.90 for intermediate, and 0.83 for good collaterals (Fig. 4). An average sensitivity of 80.33% and specificity of 79.33% were obtained by taking the mean true positive and false positive rates of the three classes. The feature ranking and classification were performed using scikit-learn [23].

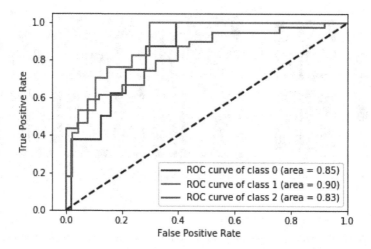

Fig. 4. ROC curve showing classification performance of good, intermediate and poor collaterals.

Further, we compared SVM performance with the Random Forest (RF). However, due to our small sample size and imbalanced dataset, RF performed extensive model selection which led to over-fitting. Thus although we had a high training accuracy (98%), the testing accuracy was 55% in the case of RF.

All the experiments were performed on a Windows 7 machine equipped with an Intel(R) Core(TM) i7-4770 CPU @ 3.40 GHz and 28 GB of RAM. The robust and automatic approach grades collaterals quickly (i.e. in approximately 10 min with 3 min for the registration step, less than 6 min for feature extraction and 1 min to classify a single patient).

4   Discussion

In the current study, we developed an automatic approach for collateral grading based on ASPECTS terminology using NCCT. This is a novel modality for assessing collaterals which are an independent predictor of good clinical outcomes. Most of the studies use dynamic CTA to assess the collateral grading whereas we proposed classifying collaterals based on ischemic damage from NCCT which requires less time, no bolus, and less radiation. To analyze the relationship between ischemic damage and collateral status, the agreement between e-CTA, which assesses collaterals with machine learning using CTA-CS [29] and e-ASPECTS from e-STROKE SUITE (Brainomix Ltd.,) is shown in the study of Grunwald *et al.* [10]. Our method uses 4D CTA as ground truth to obtain the collateral status from multiple phases and assesses it automatically from single-phase NCCT which resembles radiologists' scores and methodology.

The proposed method automatically scores collaterals using SVM based classifier which is a popular classification method used in many other ischemic stroke analysis studies and more insights can be found in the study of Kamal *et al.* [13]. Since each subject's bilateral sides resemble each other, the collateral extent can be identified from the non-symmetry of both sides. Following this idea, the difference of radiomic features between each side helps to identify ischemic damages automatically without manual intervention indicating insufficient collateral regions in the MCA territory. The atlas-based ASPECTS regions mapping to individual patient helps to improve the performance of the classifier and validate our method based on the popular ASPECTS scoring terminology.

A limitation of this study is the small dataset which is the main challenge of training a classifier. To have a reasonable ratio of train test data, we used 10-fold cross-validation. Since we have only 8 poor cases, this splitting doesn't confirm a poor case to test in each fold. In future work, we will validate our method's performance with more data before applying it to clinical trials.

In conclusion, we have implemented a machine learning-based collateral grading method using NCCT that may replace high-contrast and radiation-based CTA. By using 4D CTA based collateral scoring as ground truth, this novel approach can evaluate collaterals from the tissue degeneration extracted by radiomic features in NCCT. Our results show the feasibility of using NCCT to help physicians identify suitable patients for revascularization in AIS.

Acknowledgements. This study was funded by NSERC Discovery Grant RGPIN-2020-04612, NSERC-N01759, ENCS FRS-VE0236 and Fonds de recherche du Quebec – Nature et technologies (FRQNT Grant F01296). The author Y. Xiao is supported by BrainsCAN and CIHR fellowships.

References

1. Aspects score in acute stroke. http://aspectsinstroke.com/
2. Aktar, M., Tampieri, D., Rivaz, H., Kersten-Oertel, M., Xiao, Y.: Automatic collateral circulation scoring in ischemic stroke using 4D CT angiography with low-rank and sparse matrix decomposition. Int. J. Comput. Assisted Radiol. Surg. 1–11 (2020)
3. Bhatia, R., et al.: CT angiographic source images predict outcome and final infarct volume better than non contrast CT in proximal vascular occlusions. Stroke **42**(6), 1575–1580 (2011)

4. Burt, P., Adelson, E.: The Laplacian pyramid as a compact image code. IEEE Trans. Commun. **31**(4), 532–540 (1983)
5. Choi, J.Y., et al.: Conventional enhancement CT: a valuable tool for evaluating pial collateral flow in acute ischemic stroke. Cerebrovascular Diseases **31**(4), 346–352 (2011)
6. Christoforidis, G.A., Mohammad, Y., Kehagias, D., Avutu, B., Slivka, A.P.: Angiographic assessment of pial collaterals as a prognostic indicator following intraarterial thrombolysis for acute ischemic stroke. Am. J. Neuroradiol. **26**(7), 1789–1797 (2005)
7. Fahmi, F., Marquering, H.A., Majoie, C.B., van Walderveen, M.A., Streekstra, G.J., et al.: Image based automated aspect score for acute ischemic stroke patients. In: 2017 5th International Conference on Instrumentation, Communications, Information Technology, and Biomedical Engineering, pp. 1–5. IEEE (2017)
8. Fonov, V., et al.: Unbiased average age-appropriate atlases for pediatric studies. Neuroimage **54**(1), 313–327 (2011)
9. Grotta, J.C., et al.: Agreement and variability in the interpretation of early CT changes in stroke patients qualifying for intravenous RTPA therapy. Stroke **30**(8), 1528–1533 (1999)
10. Grunwald, I.Q., et al.: Collateral automation for triage in stroke: evaluating automated scoring of collaterals in acute stroke on computed tomography scans. Cerebrovascular Diseases **47**(5–6), 217–222 (2019)
11. Higashida, R.T., Furlan, A.J.: Trial design and reporting standards for intraarterial cerebral thrombolysis for acute ischemic stroke. Stroke **34**(8), e109–e137 (2003)
12. Jung, S., Wiest, R., Gralla, J., McKinley, R., Mattle, H., Liebeskind, D.: Relevance of the cerebral collateral circulation in ischaemic stroke: time is brain, but collaterals set the pace. Swiss medical weekly **147**(w14538), w14538 (2017)
13. Kamal, H., Lopez, V., Sheth, S.A.: Machine learning in acute ischemic stroke neuroimaging. Front. Neurol. **9**, 945 (2018)
14. Kellner, E., Reisert, M., Kiselev, V., Maurer, C., Urbach, H., Egger, K.: Comparison of automated and visual dwi aspects in acute ischemic stroke. J. Neuroradiol. **46**(5), 288–293 (2019)
15. Kersten-Oertel, M., Alamer, A., Fonov, V., Lo, B., Tampieri, D., Collins, L.: Towards a computed collateral circulation score in ischemic stroke. arXiv preprint arXiv:2001.07169, September 2016
16. Kuang, H.: Automated aspects on noncontrast CT scans in patients with acute ischemic stroke using machine learning. Am. J. Neuroradiol. **40**(1), 33–38 (2019)
17. Maas, M.B., et al.: Collateral vessels on CT angiography predict outcome in acute ischemic stroke. Stroke **40**(9), 3001–3005 (2009)
18. Maegerlein, C.: Automated calculation of the Alberta stroke program early CT score: feasibility and reliability. Radiology **291**(1), 141–148 (2019)
19. Menon, B., Smith, E., Modi, J., Patel, S., Bhatia, R., Watson, T., Hill, M., Demchuk, A., Goyal, M.: Regional leptomeningeal score on ct angiography predicts clinical and imaging outcomes in patients with acute anterior circulation occlusions. Am. J. Neuroradiol. **32**(9), 1640–1645 (2011)
20. Miteff, F., Levi, C.R., Bateman, G.A., Spratt, N., McElduff, P., Parsons, M.W.: The independent predictive utility of computed tomography angiographic collateral status in acute ischaemic stroke. Brain **132**(8), 2231–2238 (2009)
21. Mokli, Y., Pfaff, J., dos Santos, D.P., Herweh, C., Nagel, S.: Computer-aided imaging analysis in acute ischemic stroke-background and clinical applications. Neurol. Res. Pract. **1**(1), 23 (2019)

22. Muschelli, J., Ullman, N.L., Mould, W.A., Vespa, P., Hanley, D.F., Crainiceanu, C.M.: Validated automatic brain extraction of head CT images. Neuroimage **114**, 379–385 (2015)

23. Pedregosa, F., et al.: Scikit-learn: machine learning in Python. J. Mach. Learn. Res. **12**, 2825–2830 (2011)

24. Pexman, J.W., et al.: Use of the Alberta stroke program early CT score (aspects) for assessing CT scans in patients with acute stroke. Am. J. Neuroradiol. **22**(8), 1534–1542 (2001)

25. Sallustio, F., et al.: Ct angiography aspects predicts outcome much better than noncontrast CT in patients with stroke treated endovascularly. Am. J. Neuroradiol. **38**(8), 1569–1573 (2017)

26. Seker, F., Potreck, A., Möhlenbruch, M., Bendszus, M., Pham, M.: Comparison of four different collateral scores in acute ischemic stroke by CT angiography. J. Neurointervent. Surg. **8**(11), 1116–1118 (2016)

27. Shieh, Y., et al.: Computer-aided diagnosis of hyperacute stroke with thrombolysis decision support using a contralateral comparative method of CT image analysis. J. Digit. Imaging **27**(3), 392–406 (2014)

28. Sundaram, V., et al.: Automated aspects in acute ischemic stroke: a comparative analysis with CT perfusion. Am. J. Neuroradiol. **40**(12), 2033–2038 (2019)

29. Tan, I., et al.: CT angiography clot burden score and collateral score: correlation with clinical and radiologic outcomes in acute middle cerebral artery infarct. Am. J. Neuroradiol. **30**(3), 525–531 (2009)

30. Van Griethuysen, J.J., et al.: Computational radiomics system to decode the radiographic phenotype. Cancer Res. **77**(21), e104–e107 (2017)

31. Xiao, Y., et al.: Towards automatic collateral circulation score evaluation in ischemic stroke using image decompositions and support vector machines. In: Cardoso, M.J., et al. (eds.) CMMI/SWITCH/RAMBO -2017. LNCS, vol. 10555, pp. 158–167. Springer, Cham (2017). https://doi.org/10.1007/978-3-319-67564-0_16

Image-Based Subthalamic Nucleus Segmentation for Deep Brain Surgery with Electrophysiology Aided Refinement

Igor Varga[1]($\boxtimes$) (iD), Eduard Bakstein[1,3]($\boxtimes$) (iD), Greydon Gilmore[2]($\boxtimes$) (iD), and Daniel Novak[1]($\boxtimes$) (iD)

[1] Czech Technical University in Prague, 160 00 Prague, Czech Republic
{varhaiho,eduard.bakstein}@fel.cvut.cz
[2] Western University, Ontario, ON N6A 3K7, Canada
greydon.gilmore@gmail.com
[3] National Institute of Mental Health, 50 67 Klecany, Czech Republic

Abstract. Identification of subcortical structures is an essential step in surgical planning for interventions such as the deep brain stimulation (DBS), in which permanent electrode is implanted in a precisely defined location. For refinement of the target localisation and compensation of brain shift occurring during the surgery, intra-operative electrophysiological recording using microelectrodes is usually undertaken.

In this paper, we present a multimodal method that consists of a) subthalamic nucleus (STN) segmentation from magnetic resonance T2 images using 3D active contour fitting and b) a subsequent brain shift compensation step, increasing the accuracy of microelectrode placement localisation by the probabilistic electrophysiology-based fitting. The method is evaluated on a data set of 39 multi-electrode trajectories from 20 patients undergoing DBS surgery for Parkinson's disease in a leave-one-subject-out scenario. The performance comparison shows increased sensitivity and slightly decreased specificity of STN identification using the individually-segmented 3D contours, compared to electrophysiology-based refinement of a standard 3D atlas.

To achieve accurate segmentation from the low-resolution clinical T2 images, a more sophisticated approach, including shape priors and intensity model, needs to be implemented. However, the presented approach is a step towards automatic identification of microelectrode recording sites and possibly also an assistive system for the DBS surgery.

Keywords: Active contours · Deep brain stimulation · Surface fitting · Subthalamic nucleus

1 Introduction

Accurate identification of subcortical structures from medical images plays a vital role in the planning of stereotactic surgery in neurological diseases, such

© Springer Nature Switzerland AG 2020
T. Syeda-Mahmood et al. (Eds.): ML-CDS 2020/CLIP 2020, LNCS 12445, pp. 34–43, 2020.
https://doi.org/10.1007/978-3-030-60946-7_4

as the Parkinson disease (PD) or Dystonia. During the last 20 years, the DBS, targeted in the Subthalamic nucleus, has become an established treatment for late-stage treatment-resistant PD. The electrode implantation planning relies on pre-operative cranial magnetic resonance imaging (MRI) with visualisation of the target nucleus using spin-spin relaxation (T2) [8]. Further positioning of the electrode can be refined intra-operatively by the injection of exploratory electrodes and multitrajectory microelectrode-recording (MER) during surgery [4].

Currently, research studies show robust approaches to perform image segmentation of subcortical structures [12,16]. These segmentation techniques use intensity-based probabilistic models to identify individual patient's STN and support planning targets for electrode positions. However, during surgery, the target may be displaced from the planned position due to brain shift and electrode bending.

Here, we introduce a segmentation model that allows evaluation of brain shift during surgery based on the segmented STN contour and MER-based refinement. The training stage in segmentation algorithms usually requires manual labelling of the target structure volume, which requires expert knowledge and is highly time-consuming. The Active Contour Model (ACM) approaches were applied in medical image processing previously [6] as an intensity-based segmentation technique. One of our goals is to minimise labelling time, by using partial labelling of STN using just two landmark points, where the ACM will carry out adjustment of the nucleus borders based on image properties.

In the last stage, we use an electrophysiology-based model to estimate the displacement of exploratory electrodes with respect to the surgical plan and their position within STN using the model previously derived from the MRI images.

2   Methods

The approach we used can be divided into the following steps: intensity normalisation over white matter (WM), STN atlas mesh positioning and border adjustment, which results in an individual STN surface representation. A maximum-likelihood translation of this surface model is then found according to a probabilistic model of electrophysiological activity inside/outside of the nucleus.

2.1   Data

The data we used consists of 20 subjects, each of them containing:

1. Preoperative spin-lattice relaxation T1-weighted images (TE $= 1.5$ ms, TI $= 300$ ms)
2. Preoperative T2-weighted (TE $= 110$ ms, TR $= 2800$ ms) slab slices
3. Intra-operative MER recording with recording frequency 24 kHz

This group consists of subjects who underwent STN-targeted DBS therapy at the University of Western Ontario, Canada. All MRI images were recorded with a 1.5T clinical scanner (Signa 1.5T scanner, General Electric, Milwaukee,

Wisconsin, USA) two weeks prior to surgery. The slice thickness was 1.5 mm. STN was manually labelled by two landmark points: the most anterior (P1) and the most dorsal part (P2) [14].

2.2   MER Acquisition and Preprocessing

A computer-controlled microelectrode drive was mounted to the stereotactic frame (StarDrive, FHC Inc., Bowdoinham, ME), and 2–5 cannulas with tungsten microelectrodes (60 μm diameter) were lowered to 10.0 mm above the surgically planned target. Electrophysiological signals were recorded in increments of 1.00 mm (10.00 mm to 5.00 mm above the surgical target) and 0.50 mm (5.00 mm above the target until the substantia nigra reticulata was reached, marking the ventral STN border). At each recording site, data was collected for 10 s, which resulted in approximately 25–30 recordings for each microelectrode. The signals were sampled (24 kHz, 8-bit), amplified (gain: 10000) and digitally filtered (bandpass: 500–5000 Hz, notch: 60 Hz) using the Leadpoint recording station (Leadpoint 5, Medtronic).

Then, stationary segments of the electrophysiology data were first identified using the covariance method from [3], and the root-mean-square value has been calculated from the stationary segments for each signal. Next, the normalised root-mean-square values (NRMS) were calculated for each trajectory by using the first five recording positions as a reference [11]. All MER processing and fitting were performed using Matlab (MathWorks Inc., Natick, MA, USA).

2.3   MRI Data Processing

Intensity Normalisation. Due to a difference of intensity values between subjects, we performed intensity normalisation using fuzzy C means-based normalisation, which normalises the intensity of the white matter (WM) [13].

For deriving WM mask, we used the Brain extraction tool (BET) form the FSL package [15] on T1-weighted images, with subsequent use of automated segmentation tool [17]. The patient data was processed in the patient native space.

Atlas Initialisation. Initial STN was presented as an atlas 3D mesh derived from Harvard Oxford subcortical atlas [16].

For positioning the atlas mesh into the native space, we used the manually expert-labelled STN landmarks (P1 and P2) and the posterior commissure (PC).

2.4   Active Contours Fitting

For adjusting the STN boundary, we used an active contour model (ACM), which can be described as snake model [9]. Snake model implement the idea of iterative updates of a boundary curve $v(s) = (x(s), y(s), z(s))$ along the contour segment s, minimising the energy

$$E^*_{snake} = \int_0^1 E_{snake}(v(s))ds = \int_0^1 E_{int}(v(s)) + E_{image}(v(s)) + E_{con}(v(s))ds,$$

where E_{int} represents internal energy of bending spline, E_{image} is an energy represented by image forces, E_{con} represents constraints.

In the presented case of STN segmentation, where the boundary is not well defined by intensity gradient, we used a Chan-Vese model (CV) [5] whose stopping condition of curve evolution is not defined by the gradient but as the optimum of an energy function:

$$F(S, s_1, s_2) = \mu \cdot area(S) + v \cdot volume(inside S)$$
$$+ \lambda_1 \int_{inside(S)} |I - s_1|^2 dx dy dz$$
$$+ \lambda_2 \int_{outside(S)} |I - s_2|^2 dx dy dz,$$

where I is the image, S is the surface, s_1 is the average intensity inside the surface and s_2 is average intensity outside the surface. The $\mu, v, \lambda_1, \lambda_2$ are the method parameters.

As the MRI data is coarse in terms of spatial resolution (i.e. large voxel size of the 1.5T data), while on the other hand the STN model is a triangular mesh with high level of detail, we modified the original method to work in the following way: We adapted the model from [10]:

$$V(S, s_1, s_2) = |\frac{dS(u)}{dn}| \cdot (\lambda_1 \cdot |I - s_1|^2 + \lambda_2 \cdot |I - s_2|^2)$$

$$u = \begin{cases} 1 & if \quad V < 0, \text{ inside of the surface} \\ 0 & if \quad V \geq 0, \text{ outside of the surface,} \end{cases}$$

where S is optimisation surface, u is structure volume mask recalculated on each algorithm iteration, n is a vector of normal to the vertex of the mesh. We implemented this model with updating mesh vertices using gradients along vertex normals on each iteration. In each iteration, we thus updated the position of each vertex along the direction of its normal. The gradient along vertex normals was calculated using b-spline interpolation of voxel-based intensities.

After each iteration, we used a shape normalisation step, which prevents our vertices to go too close to each other and make them more uniformly distributed throughout the surface of the STN. In this constraint, we move each vertex in the direction of the largest adjacent mesh triangle, as suggested by [12].

As parameters of the CV model, we set λ_1 higher than λ_2, which made the variance of intensity inside of segmented volume lower than outside. We did not consider λ_1 and λ_2 being equal due to a marked difference in intensity values inside of STN and outside.

2.5   MER-based Fitting

In the next step, we shift the 3D STN contour, resulting from the segmentation, to fit the recorded electrophysiology. For this purpose, we use the parametric probabilistic model based on NRMS values, described previously in [1,2]. We search for a translation of the segmented STN volume (i.e. shift along the x, y and z axis) that minimises the negative log-likelihood of the atlas position with respect to the MER data:

$$t^* = arg\min_t \sum_{i=1}^{N} -\ln(p(\{x_i, l_i\}|t, \Theta))$$

where t^* is the resulting translation vector along the x, y, z axes, x_i are the N NRMS values measured at locations l_i (i.e. the MER recording sites) and Θ are parameters of the probabilistic model estimated on training data. The resulting translation t^* is then applied to the STN model and evaluated. Contrary to the original work, no scaling or rotation was done at this stage.

2.6   Evaluation Procedure

In order to evaluate the performance of the presented model, including MER-based fitting, we used an iterative leave-one-subject-out (LOSO) cross-validation. In each iteration, all (i.e. one or two) trajectories of a single patient were kept for validation, while all remaining data were used for calculating model parameters. Once the procedure was finished for all subjects, the validation set performance was evaluated. The image-based segmentation using the modified Chan-Vese algorithm was completed prior to the LOSO procedure. A summary of the process at each iteration was as follows:

1. Train the parameters of the MER-based model using MER data of the current training $(N-1)$ subjects.
2. Perform the MER-based fit on the test subject using a) the landmark-initialised anatomical atlas, or b) the result of the modified Chan-Vese algorithm.

3   Results and Discussions

Here we present and evaluate the results of the active-contour border adjustment and electrode shifting and discuss limitations and future directions of the whole pipeline.

3.1   STN Segmentation

First, we analyse volumetric properties of the landmark-initialised atlas mesh (using the P1 and P2 points) and compare it with the mesh after the Chan-Vese segmentation. See the Table 1) which compares the result with previously published STN volumetric data [18].

We can see that the initialised volume is slightly lower compared to the reference values. This differences may be due to the similar intensity values in the border between STN and Substantia nigra (SNr) and issues in labelling. Additionally, the reference values [18] may yield different results than the classical MRI intensity-based segmentation.

Table 1. STN volumes measured for initialised model by manual labels and for Chan-Vese adjusted model

	Initialised atlas, mm^3	CV segmentation, mm^3	Reference [18], mm^3
Left	101.88 ± 45.07	101.10 ± 33.05	128.8 ± 17.10
Right	76.65 ± 21.47	132.14 ± 70.95	134.52 ± 22.82

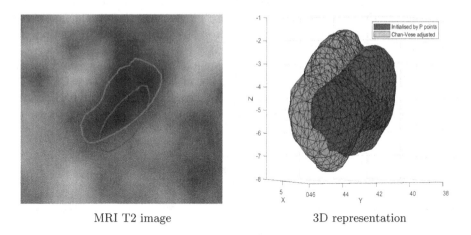

MRI T2 image   3D representation

Fig. 1. Both evaluated mesh representations: the landmark-initialised atlas (red) and the modified Chan-Vese segmentation (blue). The contours are shown overlaid over the T2 image (1a) and as a 3D representation (1b), both in the same patient (Color figure online)

For the Chan-Vese model, we observe that model fits well with the dorso-lateral STN borders (see Fig. 1a), which are the primary targets during DBS surgery [7]. However, the observed standard deviation of segmented STN volumes is significantly higher from reference volumes which is the result of low contrast in the anterior structure border. From results (see Fig. 1), we observed that CV algorithm partially include SNR into segmented volume.

To improve the segmentation results, we suggest to build a model with shape constraints and intensity modelling seems necessary, as was used in previous studies [12, 16].

3.2 MER-based Fitting

The MER-based fitting represents the brain-shift correction by estimating the most likely STN shift according to the expected and observed electrophysiological activity. For evaluation, we used the expert labels of each MER recording as inside/outside of the STN and calculated the accuracy, sensitivity and specificity of correctly including/excluding each recording. The Youden's J index ($sensitivity + specificity - 1$) was calculated due to the high-class imbalance of the dataset with most recordings being from outside the STN. Table 2 and Fig. 2 present the values achieved by both surface models at their initial positions and after the MER fitting.

As seen from the results, the MER fitting improved the fit in both cases. The highest mean accuracy was achieved by the original atlas shape after MER fit, while the Chan-Vese segmentation achieved the best sensitivity and Youden's J index. This suggests the segmented STN shape better represents the individual characteristics of the highly variable STN nucleus. However, it may also be connected to the slightly higher volumes of the Chan-Vese segmentation, as discussed above.

The Fig. 3 shows the 3D situation with microelectrode recordings and initial and final fit of the atlas for both methods. It can be seen that the MER-based fitting correctly shifted both models so that the expert-labelled STN recordings (yellow electrode segments) are inside the final volume (recording locations marked by black dots).

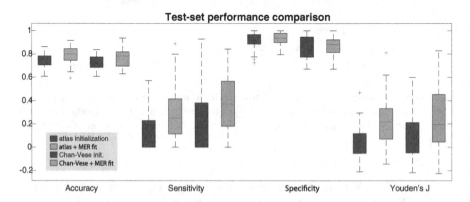

Fig. 2. Performance comparison of model location after the MER-based fitting. MER recording sites correctly included/excluded from the STN volume are considered as positive/negative examples, respectively.

Table 2. Performance evaluation of the MER-fitting results as mean (sd.), see also Fig. 2

Method	Accuracy	Sensitivity	Specificity	Youden's J
Atlas init	0.742 (0.062)	0.112 (0.151)	0.917 (0.073)	0.029 (0.136)
Atlas MERfit	0.794 (0.071)	0.283 (0.211)	0.930 (0.057)	0.214 (0.215)
CV init	0.726 (0.061)	0.233 (0.247)	0.865 (0.099)	0.098 (0.189)
CV MERfit	0.762 (0.077)	0.376 (0.241)	0.863 (0.085)	0.240 (0.239)

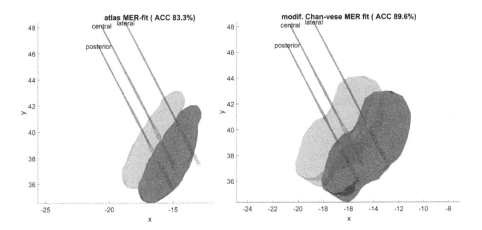

Fig. 3. MER-based fitting using the atlas surface model (left) and the modified Chan-Vese segmentation (right) for the same subject as in Fig. 1. The electrode trajectories are represented by the grey cylinders, the initial model position is shown in grey, the final one after MER fitting in violet. (Color figure online)

4   Conclusion

From the obtained results of STN segmentation, we see that the modified Chan-Vese model can fit the shape to the borders from MRI images, although it lacks in terms of shape constraints especially at the STN-SNR border, where the contrast is low. This can be improved by utilising the more sophisticated active shape and appearance models, which utilises probabilistic distribution fitting of the shape and intensity aspects of the nucleus [12,16]. The advantage of this approach is also the possibility to train several brain structures simultaneously and analyse statistical properties of shape and intensity in different positions. The other approach which could improve segmentation is training point distribution model using Deep Learning techniques which is useful in segmenting structures without prior information about other techniques.

Further, we observed that the estimation of the brain shift model could be used during surgery and allow surgeons more properly choose the target trajectory, as well as more accurately identify the MER recording sites in single-unit studies concerning STN topology and its internal structure. This was allowed by utilisation of a rare data set combining MER and MRI data from the same subjects.

A major limitation of the presented results is the lack of ground-truth STN labels in our data set, which does not enable evaluation of the actual overlap between segmented and true STN. To provide a more accurate evaluation of the fit quality, manual expert evaluation of the STN contours will be necessary.

By combining the two approaches, the preoperative MRI-based segmentation of the individual STN shape, together with intra-operative MER-based improvement, may in the future provide unprecedented accuracy in 3D MER localisation and target identification during surgery.

Acknowledgments. The study was supported by the Research Centre for Informatics, grant number CZ.02.1.01/0.0/16˜019/0000765 and by the grant Biomedical data acquisition, processing and visualisation, number SGS19/171/OHK3/3T/13. The work of EB has been supported by the Ministry of Health of the Czech Republic under the grant NV19-04-00233.

References

1. Bakštein, E., Sieger, T., Novák, D., Růžička, F., Jech, R.: Automated atlas fitting for deep brain stimulation surgery based on microelectrode neuronal recordings. In: Lhotska, L., Sukupova, L., Lacković, I., Ibbott, G.S. (eds.) World Congress on Medical Physics and Biomedical Engineering 2018. IP, vol. 68/3, pp. 105–111. Springer, Singapore (2019). https://doi.org/10.1007/978-981-10-9023-3_19

2. Bakštein, E., Sieger, T., Růžička, F., Novák, D., Jech, R.: Fusion of microelectrode neuronal recordings and MRI landmarks for automatic atlas fitting in deep brain stimulation surgery. In: Stoyanov, D., et al. (eds.) CARE/CLIP/OR 2.0/ISIC - 2018. LNCS, vol. 11041, pp. 175–183. Springer, Cham (2018). https://doi.org/10.1007/978-3-030-01201-4_19

3. Bakštein, E., et al.: Methods for automatic detection of artifacts in microelectrode recordings. J. Neurosci. Meth. **290**, 39–51 (2017)

4. Bjerknes, S., et al.: Multiple microelectrode recordings in STN-DBS surgery for Parkinson's disease: a randomized study. Mov. Disord. Clin. Pract. **5**(3), 296–305 (2018)

5. Chan, T., Vese, L.: An active contour model without edges. In: Nielsen, M., Johansen, P., Olsen, O.F., Weickert, J. (eds.) Scale-Space 1999. LNCS, vol. 1682, pp. 141–151. Springer, Heidelberg (1999). https://doi.org/10.1007/3-540-48236-9_13

6. Chen, X., Williams, B.M., Vallabhaneni, S.R., Czanner, G., Williams, R., Zheng, Y.: Learning active contour models for medical image segmentation. In: 2019 IEEE/CVF Conference on Computer Vision and Pattern Recognition (CVPR), pp. 11624–11632. IEEE, Long Beach, CA, USA, June 2019

7. Coenen, V.A., Prescher, A., Schmidt, T., Picozzi, P., Gielen, F.L.H.: What is dorsolateral in the subthalamic Nucleus (STN)?–a topographic and anatomical consideration on the ambiguous description of today's primary target for deep brain stimulation (DBS) surgery. Acta Neurochir. (Wien) **150**(11), 1163–1165 (2008)

8. Groiss, S., Wojtecki, L., Südmeyer, M., Schnitzler, A.: Review: deep brain stimulation in Parkinson's disease. Ther. Adv. Neurol. Disord. **2**(6), 379–391 (2009)
9. Kass, M., Witkin, A., Terzopoulos, D.: Snakes: active contour models. Int. J. Comput. Vision **1**(4), 321–331 (1988)
10. Marquez-Neila, P., Baumela, L., Alvarez, L.: A morphological approach to curvature-based evolution of curves and surfaces. IEEE Trans. Pattern Anal. Mach. Intell. **36**(1), 2–17 (2014)
11. Moran, A., Bar-Gad, I., Bergman, H., Israel, Z.: Real-time refinement of subthalamic nucleus targeting using Bayesian decision-making on the root mean square measure. Mov. Disord. **21**(9), 1425–1431 (2006)
12. Patenaude, B., Smith, S.M., Kennedy, D.N., Jenkinson, M.: A Bayesian model of shape and appearance for subcortical brain segmentation. NeuroImage **56**(3), 907–922 (2011)
13. Reinhold, J.C., Dewey, B.E., Carass, A., Prince, J.L.: Evaluating the impact of intensity normalization on MR image synthesis. arXiv:1812.04652 [cs], December 2018. arXiv: 1812.04652
14. Sieger, T., et al.: Distinct populations of neurons respond to emotional valence and arousal in the human subthalamic nucleus. Proc. Natl. Acad. Sci. **112**(10), 3116–3121 (2015)
15. Smith, S.M.: Fast robust automated brain extraction. Hum. Brain Mapp. **17**(3), 143–155 (2002)
16. Visser, E., Keuken, M.C., Forstmann, B.U., Jenkinson, M.: Automated segmentation of the substantia nigra, subthalamic nucleus and red nucleus in 7 T data at young and old age. Neuroimage **139**, 324–336 (2016)
17. Zhang, Y., Brady, M., Smith, S.: Segmentation of brain MR images through a hidden Markov random field model and the expectation-maximization algorithm. IEEE Trans. Med. Imaging **20**(1), 45–57 (2001)
18. Zwirner, J., et al.: Subthalamic nucleus volumes are highly consistent but decrease age-dependently-a combined magnetic resonance imaging and stereology approach in humans. Hum. Brain Mapp. **38**(2), 909–922 (2017)

3D Slicer Craniomaxillofacial Modules Support Patient-Specific Decision-Making for Personalized Healthcare in Dental Research

Jonas Bianchi[1]([⊠]), Beatriz Paniagua[2], Antonio Carlos De Oliveira Ruellas[1],
Jean-Christophe Fillion-Robin[2], Juan C. Prietro[3], João Roberto Gonçalves[4],
James Hoctor[2], Marília Yatabe[1], Martin Styner[3], TengFei Li[3], Marcela Lima Gurgel[5],
Cauby Maia Chaves Junior[5], Camila Massaro[6], Daniela Gamba Garib[6],
Lorena Vilanova[6], Jose Fernando Castanha Henriques[6], Aron Aliaga-Del Castillo[6],
Guilherme Janson[6], Laura R. Iwasaki[7], Jeffrey C. Nickel[7], Karine Evangelista[8],
and Lucia Cevidanes[1]

[1] University of Michigan, 1011 North University Ave, Ann Arbor, MI 48109, USA
bianchij@umich.edu
[2] Kitware Incorporation, Carrboro, NC 27510, USA
[3] University of North Carolina, Chapel Hill, NC 27599, USA
[4] São Paulo State University, Araraquara, SP 14801-903, Brazil
[5] Federal University of Ceara, Fortaleza, CE 60020-181, Brazil
[6] University of São Paulo, Bauru, SP 17011-220, Brazil
[7] Oregon Health & Science University, Portland, OR 97201, USA
[8] Federal University of Goias, Goiania, GO 74690-900, Brazil

Abstract. The biggest challenge to improve the diagnosis and therapies of Craniomaxillofacial conditions is to translate algorithms and software developments towards the creation of holistic patient models. A complete picture of the individual patient for treatment planning and personalized healthcare requires a compilation of clinician-friendly algorithms to provide minimally invasive diagnostic techniques with multimodal image integration and analysis. We describe here the implementation of the open-source Craniomaxillofacial module of the 3D Slicer software, as well as its clinical applications. This paper proposes data management approaches for multisource data extraction, registration, visualization, and quantification. These applications integrate medical images with clinical and biological data analytics, user studies, and other heterogeneous data.

Keywords: Personalized medicine · Data management · Data analytics · Craniomaxillofacial diseases · Dental research

1 Introduction

3D Slicer is an open-source software for medical data visualization and image analysis to establish a platform for clinical studies [1, 2], and it is available for multiple operating systems: Linux, macOS, and Windows [1–4]. It supports additional extensions/modules

© Springer Nature Switzerland AG 2020
T. Syeda-Mahmood et al. (Eds.): ML-CDS 2020/CLIP 2020, LNCS 12445, pp. 44–53, 2020.
https://doi.org/10.1007/978-3-030-60946-7_5

through three principal approaches: Command Line Interface (CLI), Loadable Modules, and Scripted Modules based on different languages (such as C++ and python), that can be built, tested, packaged, and distributed in the 3D Slicer "Extensions Manager" [5, 6]. In recent years, medical and engineer researchers focused on developing 3D Slicer modules and algorithms distributed as extensions allowing users to work with datasets from different organs, and multi-modality images, such as magnetic resonance image (MRI), computed tomography (CT), cone-beam CT (CBCT), Ultrasound (US), nuclear medicine and microscopy [6]. In this paper, we present the compilation of a 3D Slicer module called: *3D Slicer Craniomaxillofacial (CMF)* (Fig. 1), which has many extensions and applications used for dental and craniomaxillofacial research areas. Our team has developed each module individually, and we have tested the clinical research applicability by conducting user studies. Here, we also provide insights and examples of clinical research applications using the *Craniomaxillofacial* module to support patient-specific decision-making, data analysis, and visualization for personalized healthcare.

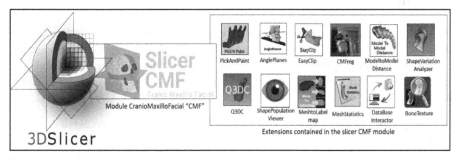

Fig. 1. Slicer Craniomaxillofacial (CMF) and its current extension. Each extension represented here has specific tasks as explored in this paper.

2 Methods

In this section, we report each extension implemented in the Craniomaxillofacial (CMF) 3D Slicer module (https://cmf.slicer.org/), developed by the Dental and Craniofacial Bionetwork for Image Analysis (DCBIA) team. The source code for the CMF 3D Slicer module can be found at https://github.com/DCBIA-OrthoLab. Table 1 describes the code availability, brief indications for user studies, references to the documentation and code of each CMF extension.

Table 1. Extensions contained in the 3D Slicer CMF module.

CMF extension	Code available at https://github.com/	Main indications
CMFreg [7]	DCBIA-OrthoLab/CMFreg	Data Registration/Processing
MeshtoLabelMap [8]	NIRALUser/MeshToLabelMap	Data Processing
AnglePlanes [9]	DCBIA-OrthoLab/AnglePlanes-Extension	Data Quantification/Visualization
Q3DC [10]	DCBIA-OrthoLab/Q3DCExtension	Data Quantification/Visualization

(*continued*)

Table 1. (*continued*)

CMF extension	Code available at https://github.com/	Main indications
Easyclip [11]	DCBIA-OrthoLab/EasyClip-Extension	Data Processing
ModeltoModelDistance [12]	NIRALUser/3DMetricTools	Data Quantification/Visualization
PickandPaint [13]	DCBIA-OrthoLab/PickAndPaintExtension	Data Management/Processing
ShapePopulationViewer [14]	NIRALUser/ShapePopulationViewer.	Data Visualization
DataBaseInteraction [15]	DCBIA-OrthoLab/DatabaseInteractorExtension	Data Management/Processing
MeshStatistics [16]	DCBIA-OrthoLab/MeshStatisticsExtension	Data Statistics
BoneTexture [17]	Kitware/BoneTextureExtension/	Data Extraction/Quantification
ShapeVariationAnalyzer [18]	DCBIA-OrthoLab/ShapeVariationAnalyzer	Data Processing/Analytics
SPHARM-PDM [19]	NIRALUser/SPHARM-PDM	Data Processing/Analytics
MFSDA [20]	DCBIA-OrthoLab/MFSDA_Python	Data Integration and Statistical Modeling

3 Results and User Studies Applications

This section reports user studies of how the extensions contained in the 3D Slicer CMF module lead to precise and personalized healthcare for the patients. The data management and analytics output patient-specific information from multisource image modalities, including CBCT scans, 3D surface digitized models, and MRIs, as well as clinical and biological information.

Patient-Specific 3D Mesh Model, and Label Map Creation. The success of craniofacial surgeries dependents on visual treatment planning in 3D, and fabrication of surgical stents to aid the precise placement of osseous structures. The ModelMaker [21] is a 3D Slicer core module that computes and creates the 3D mesh models from binary segmentations. Other image analysis steps to evaluate treatment effects or disease progression, such as voxel-based image registration and region of interest annotation, require a label map volume. For this purpose, the MeshtoLabelMap extension [8] creates a label map from a mesh VTK polydata (Fig. 2).

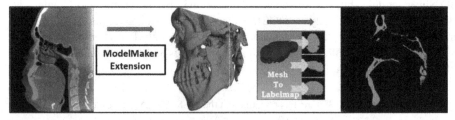

Fig. 2. Workflow for creating a 3D mesh model from the binary segmentation of a CBCT scan using ModelMaker. Then a label map volume is created using the MeshtoLabelMap Extension.

CBCT Voxel-Based Registration for Follow-up Patient Assessment. Outcomes and long-term stability of craniofacial surgery requires a means of assessment of regional skeletal changes in 3D. In this example, two CBCT scans at two time points (T2-T1) are registered using voxel-based rigid registration in the CMFreg extension [7]. The cranial base was the stable anatomical structure used as reference to assess the skeletal changes in the maxilla and mandible (Fig. 3). This extension also contains tools to downsize the voxel size, perform label extraction from the images, mask creation, surface registration, and application of matrix.

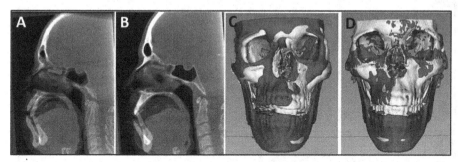

Fig. 3. Workflow with the CMFreg extension, the red arrow is pointing to the cranial base (region of registration). A- Superimposed baseline (T1) and follow-up (T2) original CBCT scans. B- Registered T1 and T2 CBCT scans. C- 3D model overlays T1 (white), and Time T2 (red) before registration. D- 3D model overlays after registration at the cranial base, note the mandibular downward and forward displacement after treatment. (Color figure online)

Quantification of Patient Craniomaxillofacial Planes, Linear and Angular Measures. Orthodontics and Orthognathic surgery diagnosis and treatment planning require quantification of structures in 3D. The AnglePlanes extension [9] was developed to quantify the specific patients' craniofacial angular measurements [22], enabling a precise diagnosis, e.g. measurement of angle between lower borders of mandible and skull (Fig. 4), and treatment plan.

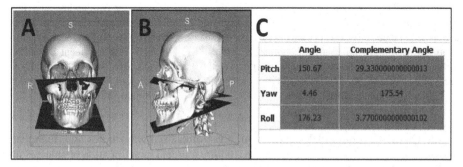

	Angle	Complementary Angle
Pitch	150.67	29.330000000000013
Yaw	4.46	175.54
Roll	176.23	3.7700000000000102

Fig. 4. A- Frontal view of the craniofacial 3D mesh model that was used to place the landmarks that defined two spatial planes for measurement of mandibular angulation. B- Lateral view of the planes. C- Pitch, Yaw, and Roll measures.

The examples shown in Fig. 5 used linear and angular measurements for patient classification and personalized assessment of changes occurred in different orthodontic treatments, such as maxillary expansion, computed with the Q3DC extension [10].

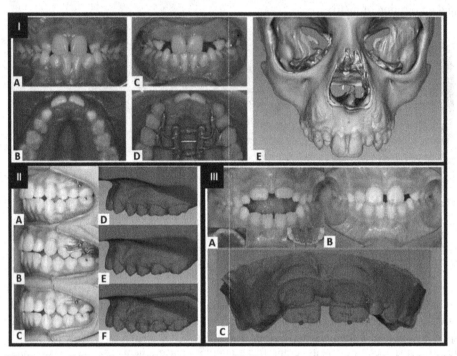

Fig. 5. Case I) Pre- (A, B) and post treatment (C, D) records of a patient treated with rapid maxillary expansion. The 3D models derived from the CBCT scans were acquired before (white color) and after (yellow color) expansion. The cranial base superposition (E) shows the maxillary transverse changes. To quantitatively assess the distance between right and left canines after treatment, Q3DC extension was used to calculate the distance between the two landmarks at the tips of the canines (E). **Case II, and III.** Quantification of treatment effectiveness in registered digital dental models (DDM) obtained with intraoral scanners. **Case II)** A- Pretreatment; B- Post treatment; C- Follow-up treatment photos; D- Pretreatment DDM; E- Post treatment DDM; F- surface registration of the DDM and landmarks placed for quantitative assessments. **Case III)** A- Pretreatment photos; B- Post treatment photos; C- Surface registration of DDM and landmarks placed for quantitative assessments in the Q3DC extension. (Color figure online)

Image Quantification and Visualization. Craniofacial growth studies and treatment outcomes analyses of surgery interventions require regional quantification of changes in the craniofacial complex. In Fig. 6 we show examples of quantification of 3D mesh models surface distances (ModeltoModelDistance extension [12]), visualization of scalar distance maps (ShapePopulationViewer extension [14], quantification of specific anatomic regions of interest (PickandPaint extension [13]) and descriptive statistics of distances in this particular region in the mesh (MeshStatistics extension [16]).

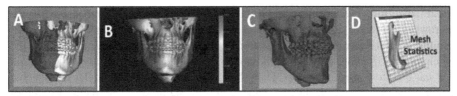

Fig. 6. Workflow for working with the four extensions. A- 3D mesh models in two time points (Red and White). B- The ModeltoModelDistance generates a VTK model that was loaded into the ShapePopulationViewer extension, where it is possible to see the colormap distances. C- The same model was loaded in the PickandPaint Extension, where the user selects a Region of Interest (ROI in red). D- Using the Extension MeshStastiscs, the user can obtain the measurements for the ROI in x, y, and z as well as the mean, SD, and percentiles. (Color figure online)

Data Extraction, Quantification, and Multimodal Image Registration. Regional analyses and longitudinal effects of disease, such as osteoarthritis of the temporomandibular joint, can be quantified by the combination of multimodal imaging data. In Fig. 7-I, we show the registration of CBCT and MR images from the same patient, using the Elastix extension [23, 24] for rigid and nonrigid registration of multimodal images (https://github.com/lassoan/SlicerElastix). In Fig. 7-II we show the data extraction from high-resolution CBCT scans, using the BoneTexture extension [17], to quantify radiomics markers, such as bone morphometry, gray-level co- occurrence, and gray level run-length matrices.

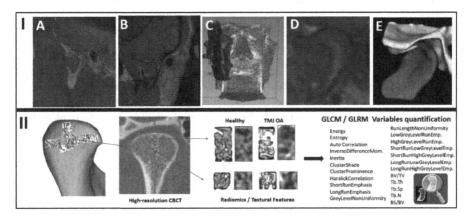

Fig. 7. I) Data Registration (I) and Radiomics Quantification (II). I) A: CBCT scan; B- MRI; C- 3D rendering of the MRI and CBCT, showing the manual approximation between the images; D- Both images registered using Elastix extension for the multimodal registration; E- 3D mesh models of the anatomical registered structures, combining the MRI and CBCT. **II)** High- resolution CBCT scan and radiomics extraction using the BoneTexture extension from healthy and TMJ OA patients. (Color figure online)

Standardization of Orientation and Region of Interest for Shape Analysis. Analyses of growth asymmetry, and degeneration of the temporomandibular joint (TMJ), requires a common orientation and standardization of the region of interest, and is one of the big challenges in image processing of population studies. Here, the Easyclip extension [11] was used to consistently crop 3D mesh models of mandibular condyles in a specific region of interest. Figure 8 shows a patient and the 3D mesh model reconstruction of the mandibular condyles from a CBCT scan. Common orientation and clipping of a standardized region of interest allow comparison of patient-specific morphology within a population.

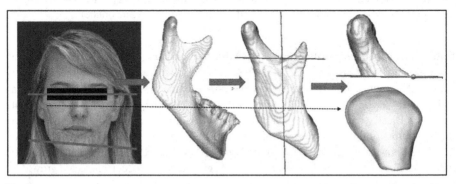

Fig. 8. Mandibular Cropping using EasyClip Extension. The mandible of this patient is asymmetric due to a unilateral condylar hyperplasia. A plane is drawn to crop the mandibular condyle for assessment of changes in this region, allowing shape comparisons among multiple time points, yellow color is T0 and white is T1 after 11 months of follow up without intervention. (Color figure online)

Imaging Data Storage, Management, and Integration. The assessment and analysis of TMJ osteoarthritis (OA), demands a comprehensive assessment of the mandibular shape and a statistical integration with clinical and biological features to explore this complex disease. In addition, the multisource biomarkers, such as molecular proteins, demographics, clinical parameters, and radiomics, require efficient data storage, management, and processing. We developed the DataBaseInteractor extension [15] to facilitate data sets from patient exams to upload directly from 3D-Slicer and export to our web system https://dsci.dent.umich.edu/. This contains several processing tools deployed, including the multivariate functional shape data analysis (MFSDA) algorithm [20] for statistical shape analysis that integrates the condylar shape and multi- source features. Figure 9, shows a case study where we integrated and correlated the clinical, biological, and radiomics features/information with the condylar shape using the MFSDA. In addition, previous to running the imaging steps described, we assessed the mandibular condylar shape correspondence using the SPHARM-PDM [19, 25] and we trained deep learning algorithms via ShapeVariationAnalyzer [18] to classify the condyles according to the level of degeneration.

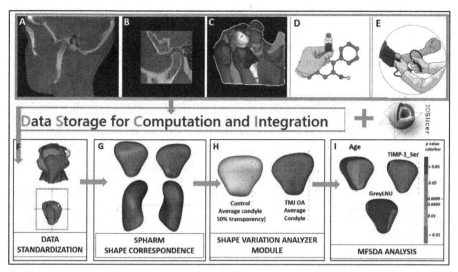

Fig. 9. Data management, storage and processing using our websystem https://dsci.dent.umich. edu/ using the DatabaseInteractor in the 3D Slicer CMF module. A- Conventional CBCT; B- High Resolution CBCT used to extract radiomics features; C- Multiple mandibular condyles from different control and TMJ OA patients; D- Illustration of protein collection; E- Illustration of clinical parameters and demographics; F- Condyles being cropped in order to have a standardized protocol for shape assessment; G- Example of a condyle after the shape correspondence using the SPHARM-PDM; H- Computation of an average mandibular condyle based on the mesh of 46 mandibular condyles in each group (control-blue and disease-red) using the SPHARM-PDM and ShapeVariationAnalyzerExtension; I- Multivariate functional shape data analysis (MFSDA) of multisource biomarkers, the color map are showing that age (demographic feature), GreyLNU (radiomics feature) and TIMP-1_serum (protein from serum) were correlated with the condylar shape in specific regions (see p-value colormap). (Color figure online)

4   Conclusion

In this manuscript, we presented the 3D Slicer Craniomaxillofacial (CMF) module and its extensions applied in clinical cases. We demonstrated the use of the different extensions to translate computational algorithms into clinical applications. This was accomplished through data processing, visualization, extraction, and combination of heterogeneous data such as radiomics, 3D mesh models, volumetric images, molecular, and clinical data.

The presented examples illustrate how the CMF module may support patient-specific decisions, with personalized treatment choices based on data management, analytics, and integration. In addition, extensions such as the MFSDA combine multisource data to facilitate feature selection and statistical modeling. The overall objective is to provide better diagnostics, information for decision-making regarding therapies, and outcomes analyses for the improved care of patients with craniomaxillofacial conditions.

Acknowledgements. This study was supported by NIH grants DE R01DE024450, R21DE025306 and R01 EB021391.

References

1. Pieper, S., Halle, M., Kikinis, R.: 3D slicer. In: 2004 2nd IEEE International Symposium Biomedical Imaging Macro to Nano, vol. 1, pp. 632–635 (2004). https://doi.org/10.1109/isbi.2004.1398617
2. Fedorov, A.A., et al.: 3D slicers as an image computing platform for the quantitative imaging network. Magn. Reson. Imaging **30**, 1323–1341 (2012). https://doi.org/10.1016/j.mri.2012.05.001.3D
3. Enquobahrie, A., Bowers, M., Ibanez, L., Finet, J., Audette, M., Kolasny, A.: Enabling ITK-based processing and 3D slicer MRML scene management in paraview, 1–10 (2012)
4. Irimia, A., Chambers, M.C., Horn, J.D. Van, Angeles, L.: Automatic segmentation of traumatic brain injury MRI volumes using atlas based classification and 3D slicer (2011)
5. Documentation/Nightly/Developers/Modules - Slicer Wiki. https://www.slicer.org/wiki/Documentation/Nightly/Developers/Modules. Accessed 03 July 2020
6. Extensions — 3D Slicer documentation. https://slicer.readthedocs.io/en/latest/developer_guide/extensions.html. Accessed 29 June 2020
7. Documentation/4.10/Extensions/CMFreg - Slicer Wiki. https://www.slicer.org/wiki/Documentation/4.10/Extensions/CMFreg. Accessed 30 June 2020
8. Documentation/4.10/Extensions/MeshToLabelMap - Slicer Wiki. https://www.slicer.org/wiki/Documentation/4.10/Extensions/MeshToLabelMap. Accessed 30 June 2020
9. Documentation/4.10/Extensions/AnglePlanes - Slicer Wiki. https://www.slicer.org/wiki/Documentation/4.10/Extensions/AnglePlanes. Accessed 30 June 2020
10. Documentation/4.10/Extensions/Q3DC - Slicer Wiki. https://www.slicer.org/wiki/Documentation/4.10/Extensions/Q3DC. Accessed 30 June 2020
11. Documentation/4.10/Extensions/EasyClip - Slicer Wiki. https://www.slicer.org/wiki/Documentation/4.10/Extensions/EasyClip. Accessed 30 June 2020
12. Documentation/4.10/Extensions/ModelToModelDistance - Slicer Wiki. https://www.slicer.org/wiki/Documentation/4.10/Extensions/ModelToModelDistance. Accessed 30 June 2020
13. Documentation/4.10/Extensions/PickAndPaint - Slicer Wiki. https://www.slicer.org/wiki/Documentation/4.10/Extensions/PickAndPaint. Accessed 30 June 2020
14. Documentation/4.10/Extensions/ShapePopulationViewer - Slicer Wiki. https://www.slicer.org/wiki/Documentation/4.10/Extensions/ShapePopulationViewer. Accessed 30 June 2020
15. Documentation/Nighty/Modules/DatabaseInteractor - Slicer Wiki. https://www.slicer.org/wiki/Documentation/Nighty/Modules/DatabaseInteractor. Accessed 30 June 2020
16. Documentation/4.10/Extensions/MeshStatistics - Slicer Wiki. https://www.slicer.org/wiki/Documentation/4.10/Extensions/MeshStatistics. Accessed 30 June 2020
17. Documentation/Nighty/Modules/BoneTexture - Slicer Wiki. https://www.slicer.org/wiki/Documentation/Nighty/Modules/BoneTexture#Module_Description. Accessed 30 June 2020
18. Documentation/Nighty/Modules/ShapeVariationAnalyzer - Slicer Wiki. https://www.slicer.org/wiki/Documentation/Nighty/Modules/ShapeVariationAnalyzer. Accessed 30 June 2020
19. Documentation/Nightly/Extensions/SpharmPdm - Slicer Wiki. https://www.slicer.org/wiki/Documentation/Nightly/Extensions/SpharmPdm. Accessed 13 July 2020
20. DCBIA-OrthoLab/MFSDA_Python. https://github.com/DCBIA-OrthoLab/MFSDA_Python. Accessed 07 July 2020
21. Documentation/4.10/Modules/ModelMaker - Slicer Wiki. https://www.slicer.org/wiki/Documentation/4.10/Modules/ModelMaker. Accessed 02 July 2020
22. Yatabe, M., et al.: Challenges in measuring angles between craniofacial structures. J. Appl. Oral Sci. **27**, e20180380 (2019). https://doi.org/10.1590/1678-7757-2018-0380
23. Klein, S., Staring, M., Murphy, K., Viergever, M.A., Pluim, J.P.W.: Elastix: a toolbox for intensity-based medical image registration. IEEE Trans. Med. Imaging **29**, 196–205 (2010). https://doi.org/10.1109/TMI.2009.2035616

24. Shamonin, D.P., Bron, E.E., Lelieveldt, B.P.F., Smits, M., Klein, S., Staring, M.: Fast parallel image registration on CPU and GPU for diagnostic classification of Alzheimer's disease. Front. Neuroinform. **7**, 50 (2010). https://doi.org/10.3389/fninf.2013.00050
25. NITRC: SPHARM-PDM Toolbox: Tool/Resource Info. https://www.nitrc.org/projects/spharm-pdm. Accessed 13 July 2020

Learning Representations of Endoscopic Videos to Detect Tool Presence Without Supervision

David Z. Li[1]($\boxtimes$), Masaru Ishii[2], Russell H. Taylor[1], Gregory D. Hager[1], and Ayushi Sinha[3]

[1] Department of Computer Science, The Johns Hopkins University, Baltimore, USA
dli44@alumni.jhu.edu
[2] Johns Hopkins Medical Institutions, Baltimore, USA
[3] Laboratory for Computational Sensing and Robotics, The Johns Hopkins University, Baltimore, USA

Abstract. In this work, we explore whether it is possible to learn representations of endoscopic video frames to perform tasks such as identifying surgical tool presence without supervision. We use a maximum mean discrepancy (MMD) variational autoencoder (VAE) to learn low-dimensional latent representations of endoscopic videos and manipulate these representations to distinguish frames containing tools from those without tools. We use three different methods to manipulate these latent representations in order to predict tool presence in each frame. Our fully unsupervised methods can identify whether endoscopic video frames contain tools with average precision of 71.56, 73.93, and 76.18, respectively, comparable to supervised methods. Our code is available at https://github.com/zdavidli/tool-presence/.

Keywords: Endoscopic video · Tool presence · Representation learning · Variational autoencoder · Maximum mean discrepancy

1 Introduction

Despite the abundance of medical image data, progress in learning from such data has been impeded by the lack of labels and the difficulty in acquiring accurate labels. With increase in minimally invasive procedures [28], an increasing number of endoscopic videos (Fig. 1) are available. This can open up the opportunity for video-based surgical education and skill assessment. Prior work [18] has shown that both experts and non-experts can produce valid objective skill assessment via pairwise comparisons of surgical videos. However, watching individual videos is time consuming and tedious. Therefore, much work is being done in automating skill assessment using supervised [5] and unsupervised [4] learning. These prior methods used kinematic data from tools to learn surgical motion. However, many endoscopic procedures do not capture kinematic data.

© Springer Nature Switzerland AG 2020
T. Syeda-Mahmood et al. (Eds.): ML-CDS 2020/CLIP 2020, LNCS 12445, pp. 54–63, 2020.
https://doi.org/10.1007/978-3-030-60946-7_6

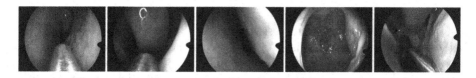

Fig. 1. Examples of endoscopic video frames with and without tools. These examples show the variance in anatomy in our dataset.

Therefore, we want to explore whether we can work towards automated skill assessment directly from endoscopic video.

Since videos contain more information than just kinematics, we want to first isolate tool motion from camera motion in endoscopic videos. If the two types of motion can be disentangled, then representations of video frames with and without tools should be distinct enough to allow separation between the two. Therefore, our aim in this work is to evaluate whether we can detect frames that contain tools. In order to do this, we use a variational autoencoder (VAE) [14] to learn latent representations of endoscopic video frames since VAEs have the ability to learn underlying low-dimensional latent representations from complex, high-dimensional input data. Specifically, we use maximum mean discrepancy (MMD) VAE [30], which uses a modified objective function in order to avoid overfitting and promote learning a more informative latent representation.

We then manipulate these learned representations or encodings using three different methods. First, we directly use the encodings produced by our MMD-VAE to evaluate whether encodings of frames with and without tools can be separated. Second, we model the tool presence as a binary latent factor and train a Bayesian mixture model to learn the clusters over our encodings and classify each frame as containing or not containing tools. Third, we use sequences of our encodings to perform future prediction and evaluate whether temporal context can better inform our prediction of tool presence. Our evaluation methods identify frames containing tools with average precision of 71.56, 73.93, and 76.18, respectively, without any explicit labels.

2   Prior Work

Prior work has shown that surgical motion can be learned from robot kinematics. Lea et al. [16] and DiPietro et al. [5] showed that supervised learning methods can accurately model and recognize surgical activities from robot kinematics. DiPietro et al. [4] further showed that encodings learned from robot kinematics in an unsupervised manner also clustered according to high-level activities. However, these methods rely on robot kinematics which provide information like gripper angle, velocity, etc. Endoscopic procedures that do not use robotic manipulation do not produce kinematics data, but do produce endoscopic videos.

Much work has also been dedicated to extracting tools from video frames using supervised tool segmentation and tool tracking methods. Many methods ignore the temporal aspect of videos and compute segmentation on a frame-by-frame basis. Several methods use established network architectures, like U-Net, to compute segmentations [20,24]. Some methods have tried to tie in the temporal aspect of videos by using recurrent neural networks (RNNs) to segment tools [1], while others have combined simpler fully convolutional networks with optical flow to capture the temporal dimension [7]. More recently, unsupervised methods for learning representations of videos have also been presented [25]. While Srivastava et al. [25] use RNNs to encode sequences of video frames, which can grow large quickly, we will explore whether unsupervised learning of video representations on a per-frame basis will give us sufficient information to discriminate between frames with and without tools.

3   Method

3.1   Dataset

We use a publicly-available sinus endoscopy video consisting of five segments of continuous endoscope movement from the front of the nasal cavity to the back [6]. The video was initially collected at 1080p resolution and split into individual frames. Frames that depicted text or where the endoscope was outside the nose were discarded. A total of 1551 frames, downsampled from 1080p resolution to a height of 64 pixels and centrally cropped to 64×64 pixels, were extracted. This downsampling was necessary due to GPU limitations.

We held out 20% of the frames, sampled from throughout our video sequence, as the test set. Each frame was manually labeled for tool presence. These annotations were used for evaluation only. In the training set, 65.8% of frames were labeled as containing a tool, and in the test set 67.4% of frames were labeled as containing a tool.

3.2   Variational Autoencoder

We use variational autoencoders (VAEs) [14] to learn low dimensional latent representations that can encode our endoscopic video data. VAEs and their extensions [30] are based on the idea that each data point, $x \in \mathbf{X}$, is generated from a d-dimensional latent random variable, $z \in \mathbb{R}^d$, with probability $p_\theta(x|z)$, where z is sampled from the prior, $p_\theta(z) \sim \mathcal{N}(\mu, \sigma^2)$, parameterized by θ [14]. However, since optimizing over the probability density function (PDF) P is intractable, the optimization is solved over a simpler PDF, Q, to find q_ϕ that best approximates p_θ [14]. To ensure that q best approximates p, θ and ϕ are jointly optimized by maximizing the evidence lower bound (ELBO) [14]:

$$\log p(x) \geq E_{q_\phi(z|x)} \left[\log p_\theta(x|z)\right] - \mathrm{KL}\left(q_\phi(z|x) \,\|\, p(z)\right). \tag{1}$$

In order to encourage VAEs to learn more informative encodings without overfitting to the data, Zhao et al. [30] introduced the maximum mean discrepancy (MMD) VAE, which maximizes the mutual information between the data and the encodings. MMD-VAE changes the objective function by replacing the KL-divergence term with MMD and introduces a regularization term, λ [30]:

$$\log p(x) \geq E_{q_\phi(z|x)} \left[\log p_\theta(x|z) \right] - \lambda \text{MMD} \left(q_\phi(z|x) \,\|\, p(z) \right). \tag{2}$$

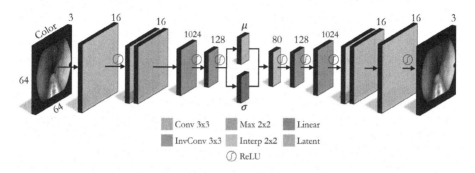

Fig. 2. Our MMD-VAE architecture with a two-layer CNN encoder and decoder.

3.3 Training

We used a convolutional neural network encoder and decoder each with two convolutional layers and three fully-connected layers (Fig. 2). We performed a hyperparameter sweep over latent dimension and regularization coefficient of our MMD-VAE implementation in PyTorch [21] and evaluated each model based on the criteria in Sect. 3.4.

The best performing model from our sweep was trained for 80 epochs using stochastic gradient descent (SGD) on a single NVIDIA Quadro K620 GPU with 2 GB memory with minibatch size of 32 and Adam optimizer [13] with default parameters except for learning rate which was set to 10^{-3}. This model had a latent dimension $d = 20$ and regularization coefficient $\lambda = 5$.

3.4 Model Evaluation

Direct Evaluation. First, we directly evaluate the encodings produced by our MMD-VAE implementation using a query-based evaluation. We compute the cosines between encodings of each test frame, i, containing a tool and all other test frames, $j \neq i$, and threshold the products to separate high and low responses. Since the query (i.e., test frame, i) contains a tool, all other test frames containing tools should produce high response, while those without tools should produce low response. To evaluate our results, we compute the average precision (AP) score [31] over responses from each test frame. AP summarizes the precision-recall curve by computing the weighted mean of precision values computed at each threshold, weighted by the increase in recall from the previous threshold. In simpler terms, AP computes the area under the precision-recall curve.

Approximate Inference. Next, we evaluate our encodings using approximate inference. We estimate p (tool presence | latent encoding) by modeling the space of encodings as a finite mixture model with a categorical latent variable C which has $K = 2$ Gaussian states (tool present and not present). We use Markov chain Monte Carlo (MCMC) sampling [19] to approximate the posterior distribution by constructing a Markov chain whose states are assignments of the model parameters and whose stationary distribution is p:

$$p(z_i|\boldsymbol{\theta}, \mu_{c_i}, \sigma_{c_i}) = \sum_{c_i=1}^{K} \boldsymbol{\theta}_{c_i}^{\top} \mathcal{N}(z_i|\mu_{c_i}, \sigma_{c_i}). \tag{3}$$

Here, $z_i \in \mathbb{R}^d$ is the ith encoding generated from our MMD-VAE, $d \in \mathbb{N}$ is the dimension of the encoding sample, and each of the K configurations are normally distributed according to parameters $\boldsymbol{\mu}, \boldsymbol{\sigma} \in \mathbb{R}^K$ and mixing probabilities $\boldsymbol{\theta} \in \mathbb{R}^{d \times K}$. We assume each encoding z_i is generated by $\boldsymbol{\theta}$ and a latent state $1 \leq c_i \leq K$, described by $\mathcal{N}(\mu_{c_i}, \sigma_{c_i})$. By running the Markov chain for B burn-in steps, we reach the stationary distribution, p.

We then sample the chain for N iterations which form samples from p [19]. The parameters of the mixture model, $\boldsymbol{\mu}, \boldsymbol{\sigma}$, and $\boldsymbol{\theta}$, are learned using the No-U-turn sampler [10] on Eq. 3. Finally, for a fixed z_i, we can estimate the probability that encoding z_i comes from latent cluster c_i to predict tool presence: $p(c_i|z_i) \propto p(z_i|c_i)p(c_i) = p(z_i, c_i)$.

For evaluation, we learned the parameters in PyStan [26] with four Markov chains with $B = N = 2500$ and default hyperparameters. The posterior probability of each sample belonging to each cluster was then computed and used to predict labels. We evaluate the separation between the two latent states and compute the AP score for our label predictions.

Future Prediction. Finally, we evaluate our encodings by training a future prediction model using a recurrent neural network (RNN) encoder-decoder [2] to observe a sequence of past video frame encodings and reconstruct a sequence of future frame encodings [4]. The intuition behind this approach is that models capable of future prediction must encode contextually relevant information [4]. Both the encoder and decoder have long short-term memory (LSTM) [8,9] architectures to avoid the vanishing gradient problem, and each frame of the future sequence is associated with its own mixture of multivariate Gaussians in order not to blur distinct futures together under a unimodal Gaussian [4].

Our PyTorch [21] implementation of the future prediction model was similar to that presented by DiPietro et al. [4]. We used 5 frame sequences of past and future encodings, and Adam [13] for optimization at a learning rate of 0.005 and other hyperparameters at their default values. The latent dimension was set to 64, the number of Gaussian mixture components to 16, and the model was trained for 1000 epochs with a batch size of 50.

As in direct evaluation, we evaluate the encodings produced by future prediction by computing the cosines between encodings from each test sequence,

s_i, containing a tool and all other test sequences, $s_j \neq s_i$, taking the maximum per-frame, and thresholding, as before, to separate high and low responses. A per-frame maximum is computed here since each frame belongs to multiple adjacent sequences, s_j, producing multiple responses. Specifically, since we used 5 frame sequences, each frame produces 5 responses, of which we pick the maximum. Finally, we compute the AP over responses from each test sequence.

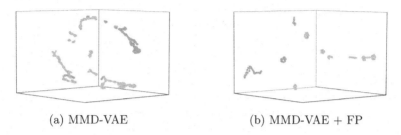

(a) MMD-VAE   (b) MMD-VAE + FP

Fig. 3. 3D dimensionality reductions, obtained using t-SNE, of (a) 80D encodings produced by MMD-VAE, and (b) 64D encodings produced by MMD-VAE + FP. MMD-VAE + FP shows slightly better separation between tool (green) and no tool (orange). The labels are used for visualization only. (Color figure online)

4   Results

The results from our three experiments are described in this section. Here, the direct evaluation will be referred to by MMD-VAE, approximate inference method by MMD-VAE + MCMC, and future prediction by MMD-VAE + FP.

We also compare our unsupervised methods to frame-level tool presence predictions from 4 *supervised* methods. Twinanda et al. [29] use a supervised CNN based on AlexNet [15] to perform tool presence detection in a multi-task manner. Sahu et al. [23] use a transfer learning approach to combine ImageNet [3] features with time-series analysis to detect tool presence. Raju et al. [22] combine features from GoogleNet [27] and VGGNet [12] for tool presence detection. Jin et al. [11] use a supervised region-based CNN to spatially localize tools and use these detections to drive frame-level tool presence detection. Results are summarized in Table 1.

Direct Evaluation. The encodings produced by our implementation of MMD–VAE show some amount of separation (Fig. 3a). Therefore, we expect our queries to produce high response when evaluated against frames with tools. However, we also expect queries to produce higher response against nearby frames with tools compared to frames that are further away from the queries. This is because our per frame encodings may not be able to disentangle the presence and absence of tools over the varied anatomy present in our video sequences (Fig. 1). Direct evaluation achieves an AP of 71.56 in identifying frames with tools.

Table 1. Average Precision (AP) in frame-level detection of tool presence
(* indicates supervised method)

Twinanda et al.*	52.5
Sahu et al.*	54.5
Raju et al.*	63.7
Jin et al.*	81.8
MMD-VAE	71.56
MMD-VAE + MCMC	73.93
MMD-VAE + FP	76.18

Approximate Inference. Since the encodings from the MMD–VAE show some
separation, we expect a trained mixture of two Gaussians to capture the separation in the encodings and generalize to labeling our held-out test set. We found
that the trained mixture model learned two clusters with distinct means and
no overlap (Fig. 4). By treating each cluster as a binary tool indicator, we can
predict labels for the encodings in test set. Compared to the direct evaluation
method, we show improvement with an AP of 73.93. We hypothesize that our
two assumptions that (1) the data can be represented as a mixture of two Gaussians, and (2) a sample belonging to a hidden configuration directly indicates tool
presence or absence may be too strong and, therefore, limiting the improvement
in AP. Instead, the clusters likely capture a combination of tool presence and
anatomy variance, and precision could be further improved by either relaxing
the assumptions or increasing the model complexity.

Future Prediction. We expect the addition of temporal information to allow
for better disentanglement between tool motion from camera motion. We observe
this improvement in the slightly greater separation in the encodings produced

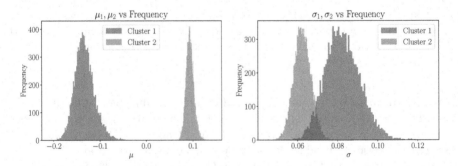

Fig. 4. Latent cluster parameter summaries for trained MMD-VAE + MCMC model
with means (left) and standard deviations (right). The two configurations of C are
described by $\mathcal{N}(-0.13, 0.0068)$ and $\mathcal{N}(0.093, 0.0039)$.

by MMD-VAE + FP than those produced without FP (Fig. 3b). This translates to further improvement in AP at 76.18. We hypothesize that larger gains in AP were again limited due to the short 5 frame sequence of encodings used to train the future prediction network. Using longer sequences may allow detection of tools over larger variations in anatomy and, therefore, improve overall results.

5   Conclusion and Future Work

We showed through our evaluations that it is possible to learn representations of endoscopic videos that allow us to identify surgical tool presence without supervision. We are able to detect frames containing tools directly from MMD–VAE encodings with an AP of 71.56. By performing approximate inference on these encodings, we are able to improve the AP of frame-level tool presence detection to 73.93. Finally, using the MMD–VAE encodings to perform future prediction allows us to further improve our AP to 76.18.

These results are comparable to those achieved by prior supervised methods evaluated on the M2cai16-tool dataset [29]. This dataset consists of 15 videos of cholecystectomy procedure, where each frame is labeled with the presence or absence of seven possible surgical tools in a multi-label fashion. As this work was evaluated on a different dataset, our immediate next step is to re-evaluate our unsupervised method on the M2cai16-tool dataset. This comparison will not only allow us to better understand how our methods compare against supervised methods, but also allow us to evaluate whether our methods can learn generalized representations across various surgical tools.

Going forward, we will explore whether variations in latent dimension and regularization for training our MMD-VAE can improve our ability to discriminate between frames with and without tools. We will also explore whether classification can be improved by accommodating the variance in anatomy by relaxing the assumption that our encodings are a mixture of two Gaussian states. Another space to explore will be whether larger sequences of video frame encodings allow us to better separate tool motion from camera motion. Although our initial work is on a limited endoscopic video dataset, our results are promising and our method can be easily applied to larger datasets with wider range of tools and anatomy since we do not rely on labels for training.

The ability to reliably identify frames containing tools can help the annotation process and can also enable further research in many different areas. For instance, methods that rely on endoscopic video frames without tools [17] can easily discard frames that are labeled as containing tools. Further, by treating features like optical flow vectors from sequences of frames with and without tools differently, we can work on identifying pixels containing tools without supervision. Unsupervised segmentation of tools, in turn, can enable unsupervised tool tracking and can have great impact on research toward video-based surgical activity recognition and skill assessment.

Acknowledgements. This work was supported by the Johns Hopkins University Provost's Postdoctoral fellowship, NVIDIA GPU grant, and other Johns Hopkins University internal funds. We would also like to thank Daniel Malinsky and Robert DiPietro for their invaluable feedback. We would also like to acknowledge the JHU Department of Computer Science providing a research GPU cluster.

References

1. Attia, M., Hossny, M., Nahavandi, S., Asadi, H.: Surgical tool segmentation using a hybrid deep CNN-RNN auto encoder-decoder. In: 2017 IEEE International Conference on Systems, Man, and Cybernetics (SMC), pp. 3373–3378, October 2017
2. Cho, K., et al.: Learning phrase representations using RNN encoder-decoder for statistical machine translation. In: Proceedings of the Conference on Empirical Methods in Natural Language Processing (EMNLP), pp. 1724–1734 (2014)
3. Deng, J., Dong, W., Socher, R., Li, L., Kai, L., Li, F.-F.: Imagenet: a large-scale hierarchical image database. In: 2009 IEEE Conference on Computer Vision and Pattern Recognition, pp. 248–255, June 2009. https://doi.org/10.1109/CVPR.2009.5206848
4. DiPietro, R., Hager, G.D.: Unsupervised learning for surgical motion by learning to predict the future. In: Frangi, A.F., Schnabel, J.A., Davatzikos, C., Alberola-López, C., Fichtinger, G. (eds.) MICCAI 2018. LNCS, vol. 11073, pp. 281–288. Springer, Cham (2018). https://doi.org/10.1007/978-3-030-00937-3_33
5. DiPietro, R., et al.: Recognizing surgical activities with recurrent neural networks. In: Medical Image Computing & Computer-Assisted Intervention, pp. 551–558 (2016)
6. Ephrat, M.: Acute sinusitis in HD (2013). www.youtube.com/watch?v=6niL7Poc_qQ
7. García-Peraza-Herrera, L.C., et al.: Real-time segmentation of non-rigid surgical tools based on deep learning and tracking. In: Computer-Assisted and Robotic Endoscopy (CARE), pp. 84–95 (2017)
8. Gers, F.A., Schmidhuber, J., Cummins, F.A.: Learning to forget: continual prediction with LSTM. Neural Comput. **12**, 2451–2471 (2000)
9. Hochreiter, S., Schmidhuber, J.: Long short-term memory. Neural Comput. **9**(8), 1735–1780 (1997)
10. Hoffman, M.D., Gelman, A.: The No-U-turn sampler: adaptively setting path lengths in hamiltonian monte carlo. J. Mach. Learn. Res. **15**(1), 1593–1623 (2014)
11. Jin, A., Yeung, S., Jopling, J., Krause, J., Azagury, D., Milstein, A., Fei-Fei, L.: Tool detection and operative skill assessment in surgical videos using region-based convolutional neural networks. In: IEEE Winter Conference on Applications of Computer Vision (2018)
12. Karen Simonyan, A.Z.: Very deep convolutional networks for large-scale image recognition. ArXiv abs/1409.1556 (2014)
13. Kingma, D.P., Ba, J.: Adam: A method for stochastic optimization. arXiv:1412.6980 (2014)
14. Kingma, D.P., Welling, M.: Auto-Encoding Variational Bayes. arXiv:1312.6114 (2013)

15. Krizhevsky, A., Sutskever, I., Hinton, G.E.: Imagenet classification with deep convolutional neural networks. In: Pereira, F., Burges, C.J.C., Bottou, L., Weinberger, K.Q. (eds.) Advances in Neural Information Processing Systems, vol. 25, pp. 1097–1105. Curran Associates, Inc. (2012). http://papers.nips.cc/paper/4824-imagenet-classification-with-deep-convolutional-neural-networks.pdf
16. Lea, C., Vidal, R., Hager, G.D.: Learning convolutional action primitives for fine-grained action recognition. In: 2016 IEEE International Conference on Robotics and Automation (ICRA), pp. 1642–1649, May 2016
17. Liu, X., et al.: Self-supervised learning for dense depth estimation in monocular endoscopy. In: Computer Assisted Robotic Endoscopy (CARE), pp. 128–138 (2018)
18. Malpani, A., Vedula, S.S., Chen, C.C.G., Hager, G.D.: A study of crowdsourced segment-level surgical skill assessment using pairwise rankings. Int. J. Comput. Assisted Radiol. Surg. **10**(9), 1435–1447 (2015). https://doi.org/10.1007/s11548-015-1238-6
19. Murphy, K.P.: Machine Learning: A Probabilistic Perspective. MIT Press, Cambridge (2012)
20. Pakhomov, D., Premachandran, V., Allan, M., Azizian, M., Navab, N.: Deep Residual Learning for Instrument Segmentation in Robotic Surgery. arXiv:1703.08580 (2017)
21. Paszke, A., et al.: Automatic differentiation in pytorch. In: NIPS-W (2017)
22. Raju, A., Wang, S., Huang, J.: M2cai surgical tool detection challenge report (2016)
23. Sahu, M., Mukhopadhyay, A., Szengel, A., Zachow, S.: Tool and phase recognition using contextual CNN features. ArXiv abs/1610.08854 (2016)
24. Shvets, A.A., Rakhlin, A., Kalinin, A.A., Iglovikov, V.I.: Automatic instrument segmentation in robot-assisted surgery using deep learning. In: 17th IEEE International Conference on Machine Learning and Applications (ICMLA), pp. 624–628 (2018)
25. Srivastava, N., Mansimov, E., Salakhutdinov, R.: Unsupervised learning of video representations using LSTMS. In: Proceedings 32nd International Conference on International Conference on Machine Learning. ICML 2015, vol. 37, pp. 843–852. JMLR.org (2015)
26. Stan Development Team: PyStan: the Python interface to Stan, Version 2.17.1.0. (2018). http://mc-stan.org
27. Szegedy, C., et al.: Going deeper with convolutions. In: Computer Vision and Pattern Recognition (CVPR) (2015). http://arxiv.org/abs/1409.4842
28. Tsui, C., Klein, R., Garabrant, M.: Minimally invasive surgery: national trends in adoption and future directions for hospital strategy. Surgical Endoscopy **27**(7), 2253–2257 (2013)
29. Twinanda, A.P., Shehata, S., Mutter, D., Marescaux, J., de Mathelin, M., Padoy, N.: Endonet: a deep architecture for recognition tasks on laparoscopic videos. IEEE Trans. Med. Imag. **36**, 86–97 (2016)
30. Zhao, S., Song, J., Ermon, S.: InfoVAE: Information Maximizing Variational Autoencoders. arXiv:1706.02262 (2017)
31. Zhu, M.: Recall, precision and average precision. In: Department of Statistics and Actuarial Science, University of Waterloo, Waterloo **2**, p. 30 (2004)

Single-Shot Deep Volumetric Regression for Mobile Medical Augmented Reality

Florian Karner[1,2], Christina Gsaxner[1,2,3], Antonio Pepe[1,2], Jianning Li[1,2], Philipp Fleck[1], Clemens Arth[1], Jürgen Wallner[2,3], and Jan Egger[1,2,3(✉)]

[1] Institute of Computer Graphics and Vision, Graz University of Technology, Graz, Austria
egger@tugraz.at
[2] Computer Algorithms for Medicine Laboratory (Café-Lab), Graz, Austria
[3] Department of Oral and Maxillofacial Surgery, Medical University of Graz, Graz, Austria

Abstract. Augmented reality for medical applications allows physicians to obtain an inside view into the patient without surgery. In this context, we present an augmented reality application running on a standard smartphone or tablet computer, providing visualizations of medical image data, overlaid with the patient, in a video see-through fashion. Our system is based on the registration of medical imaging data to the patient using a single 2D photograph of the patient. From this image, a 3D model of the patient's face is reconstructed using a convolutional neural network, to which a pre-operative CT scan is automatically registered. For efficient processing, this is performed on a server PC. Finally, anatomical and pathological information is sent back to the mobile device and can be displayed, accurately registered with the live patient, on the screen. Hence, our cost-effective, markerless approach needs only a smartphone and a server PC for image processing. We present a qualitative and quantitative evaluation using real patient photos and CT from the clinical routine in facial surgery, reporting overall processing times and registration errors.

Keywords: Augmented Reality · 3D face reconstruction · 3D registration · Deep learning · Volumetric Regression Network · Handheld devices · Smartphone · Tablet

1 Introduction

In the last years the use of digital imaging tools to support clinical diagnosis and treatment pathways have undergone a remarkable rate of technological revolution in many medical fields. This is especially true for the cranio-maxillofacial and head and neck complex where computer-assisted technologies such as automated segmentation tools, digital software packages for three dimensional (3D) preoperative visualization and operation planning or 3D printed templates of complex facial bone structures used for surgical implant adaption or others

© Springer Nature Switzerland AG 2020
T. Syeda-Mahmood et al. (Eds.): ML-CDS 2020/CLIP 2020, LNCS 12445, pp. 64–74, 2020.
https://doi.org/10.1007/978-3-030-60946-7_7

have become the goldstandard in big clinical centers [13,27,28]. These computer assisted technologies mostly work on the basis of routinely performed computed tomography (CT) or on cone beam CT scans which are daily performed in each clinical center [19,20].

In cranio-maxillofacial surgery, digital imaging data acquisition and especially CT scans are one of the standard imaging tools and are increasingly used in today's clinical routine to support clinical diagnosis and treatment plans. The usage of routinely performed digital imaging data sets such as CT scans, has increased in the last years, because of the increased digital data storage in combination with a fast data access in the clinical centers, the improved resolution that comes with new image scanner generations, resulting in more accurate anatomical imaging data, and the simultaneously reduced imaging acquisition time clinically needed for medical image data creation [17]. Therefore, software programs for processing medical image data have become a central tool in clinical medicine and are subject to continuous ongoing technological developments [26].

Augmented Reality (AR) presents an interesting opportunity to display medical imaging data, as they have been shown to provide a more intuitive interface for visualization than the standardized 2D slice view [24]. In medical AR systems, the registration of image data to the patient in the physician's view is the integral part of the technology. A straight forward approach to this problem is to use a fiducial-based method, where markers are rigidly attached to the patient [5]. These markers are tracked, either in an intrinsic fashion by the camera using computer vision methods [11,12], or extrinsically using a tracking system [2,16,18]. However, this requires excessive preparation as well as calibration, and might cause discomfort to the patient. As an alternative, surface registration algorithms present a more natural solution to the image-to-patient registration problem [15]. For obtaining surface information of the patient in this context, several methods, like depth sensors [6,7,21,22,25] or stereo cameras [1,29], are commonly used.

Instead, we propose to use an adapted convolutional neural network, called Volumetric Regression Network (VRN) by Jackson et al. [10], to reconstruct a 3D model of the patient's face from a single 2D patient photograph. We automatically register pre-interventional, volumetric imaging data to this 3D model, which enables a video see-through AR application running on a mobile device, such as a smartphone or tablet computer.

2   Materials and Methods

Our goal was to build an application for image-to-patient registration using only a single 2D facial image and a CT scan from a patient. We built a web-based client/server system by extending the existing StudierFenster framework (http://studierfenster.tugraz.at) [30,31]. The user sends data to the server via a mobile application. After that, the reconstruction and registration takes place on the server. The result of the registration step is sent back to the mobile app,

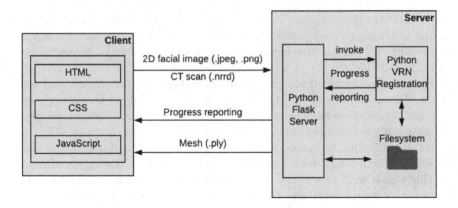

Fig. 1. Overview of the proposed client/server architecture.

where an augmented view of the patient is displayed accordingly. Figure 1 shows the proposed client/server architecture.

2.1 Data Acquisition and Preprocessing

We collect volumetric imaging data from the head and neck area of patients from clinical routine [8]. The purpose of this data is two-fold: First, it contains medical information to display in the AR environment. Second, it provides an accurate 3D model of the patients skin surface, which we exploit for image-to-patient registration. For visualization, structures of interest, such as the skull, are segmented from the imaging data using semi-automatic or manual methods. Then, a Marching Cubes algorithm [14] is applied to extract meshes of these structures. Furthermore, the skin surface is segmented using a simple thresholding approach, and a point cloud representation of the skin surface is created, which is later used for registration to the 3D model of the patient. We transform all 3D models to have their coordinate origin on the tip of the nose, which is simply determined by choosing the point with the largest z coordinate. Finally, the data is deployed to our server.

2.2 Android Application

We developed our mobile application for Android devices using the Unity 3D game engine in combination with the ARCore AR platform. Our application enables two modes, between which the user can switch. One mode is working with the front camera of the device, resulting in something like a "magic mirror" system. The second mode works with the physical back camera of the phone. Depending on the chosen camera, our application follows different pipelines to obtain an AR overlay, as shown in Fig. 2: For the front camera, we use facial detection and tracking for obtaining an augmented view of the patient, for the back camera, we make use of our proposed single-shot pipeline. The reason for these two modes is, that the front cameras have in general no simultaneous localization and mapping (SLAM) system available, because most SLAM use cases (marker tracking, 3D models, etc.) and applications (like IKEA Place), and games (like Pokémon GO) make no sense for the front camera. The front camera is mostly used for selfies and video chatting (Snapchat, FaceTime, etc.). Hence, AR SDKs do not provide SLAM systems for the front camera at all. However, a SLAM system, like it is available for the back cameras, makes an AR registration much more precise, compared to simple 2D face detection/tracking, like its implemented for the front cameras, because of the precise tracking in three-dimensional environments (continuous tracking of the camera pose and environment extracting features points for calculation of 3D world points).

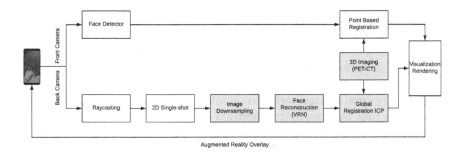

Fig. 2. Workflow of our augmented reality application. Depending on the chosen camera (front or back), our system follows different pipelines.

Front Camera Pipeline. The front camera uses ARCore's Augmented Faces module. First, a request is sent to the server to deliver the meshes for visualization and registration of the current patient. ARCore detects a person's face and estimates a 3D mesh of the facial surface, characterized by three landmarks: left forehead, right forehead and tip of the nose. We anchor the meshes received from the server at the nose tip detected by ARCore to obtain an AR overlay of the patient with virtual content.

Back Camera Pipeline. For the back camera, our proposed single-shot method is employed. To enable the placement of virtual content in a common world reference frame, we use the environmental understanding and concurrent odometry and mapping capabilities of ARCore. First, our application maps the environment around the patient using simultaneous localization and mapping (SLAM) implemented in ARCore. The user creates a 2D photograph of the patient, which is sent to the server. Simultaneously, we perform a raycast on the created environmental map through the patient's nose tip and save the hit point, which will later act as an anchor for virtual content. On the server side, 2D to 3D reconstruction and registration is performed, which will be explained in greater detail in the following section. The transformed meshes are sent back to our application, and, since they are also centered around the nose tip, we can use the hit point obtained from raycasting to place them in the AR environment accordingly.

2.3 Server Backend

To keep the computational load on the mobile device small, we perform heavy computations on a server PC. The application running on the server requires as input a frontal 2D photo of the patient, which is sent by the client, as well as 3D models from pre-interventional imaging obtained in step Sect. 2.1, which are already stored on the server.

3D Face Reconstruction Using VRN. To reconstruct a 3D model of the patient's face from a single image, we used a Convolutional Neural Network (CNN) developed by Jackson et al. [10], denoted VRN. Contrary to other solutions to the 3D face reconstruction problem, VRN does not require multiple facial images, but instead works on a single 2D image. It is able to reconstruct the entire facial geometry from a large variety of input poses and facial expressions. VRN performs this task by directly regressing a complete 3D volume using a CNN with an hourglass-like architecture, without the need to fit a 3D model to the input. Therefore, it produces fast and reliable outputs. A RGB photograph of the patient, captured with a mobile device running our application, is sent to the server and serves as input to VRN. The image has to be downsampled and rescaled to fit with the input size expected by the VRN. The output of the network is a 3D reconstructed mesh of the patient's face, from which we extract a point cloud for surface registration. Since the mesh obtained from VRN has an arbitrary unit of scale - we scale it to align it with the scale of pre-interventional imaging, which is acquired in millimeters (mm).

Surface Registration. To register the 3D model of the patients face reconstructed by VRN with the pre-interventional imaging data of the patient, we use a global registration method, followed by a refinement stage using iterative closest point (ICP), as proposed by Holz et al. [9]. For global registration, we compute fast point feature histograms [23] in both point clouds and match them iteratively using a random sample consensus algorithm [4]. This results in a coarse alignment of point clouds, which we refine by using point-to-plane ICP [3].

Mesh Transmission. The result of the surface registration is a 4 × 4 transformation matrix. With this matrix, content in the CT coordinate frame, such as an anatomical structure mesh (e.g., the bones of the skull) or pathological structure meshes (e.g., tumors) are transformed and sent back to the client. The meshes are transmitted by HTTP messages.

3   Results

We evaluated our system with ten medical scans from human subjects, seven CT scans and three magnetic resonance imaging (MRI) scans. The preoperative CT scans are from head and neck cancer patients from the clinical routine. Besides the CT, we collected frontal, high-resolution photographs of the patients, which are routinely taken before the facial operations by our clinical partners. These photos served as input to the VRN for the reconstruction and registration with the corresponding CT. In addition, we used several MRI scans of healthy subjects for testing.

3.1   Quantitative Evaluation

To quantitatively evaluate the registration error between the 3D reconstructed patient photograph and pre-interventional imaging data, we calculate the closest point registration error (CPRE) between the two point clouds. CPRE measures the average distance between points in a reference point cloud $\mathbf{P}_u^R$ with points $u = 1, 2, ..., N_R$ to their nearest points in a model point cloud $\mathbf{P}_v^M$ with $v = 1, 2, ..., N_M$, which is registered to the reference using transformation T:

$$CPRE = \frac{1}{N_R} \sum_{u=1}^{N_R} \min_{v \in [1, N_M]} ||\mathbf{P}_u^R - T \cdot \mathbf{P}_v^M||. \qquad (1)$$

Table 1 presents the total CPRE, as well as CPRE in x, y and z directions separately. Additionally, the times our system needs from end-to-end are presented. The overall time measures the following steps: (1) sending a 2D photo to the server backend, (2) performing 2D to 3D reconstruction, (3) registration of the 3D photo to the pre-operative CT and (4) transmission of registration results back to the mobile application. The measurements were repeated five times and the average was determined.

3.2   Qualitative Evaluation

For qualitative analysis, Fig. 3 displays the reconstruction and registration results of three patients. The first row shows the surface extraction from the patient's pre-operative CT. The second row shows the 3D reconstructed patient photos (anonymized with a black bar in the eye area due to patient privacy). The bottom row shows the results of the automatic registration between the surface extracted from CT (first row) and the 3D photo reconstruction (second row).

We tested our application on three healthy humans to simulate the use case with live people. All three test subjects had an MRI scan done beforehand, from which we extracted point cloud information for registration, as well as anatomical information for visualization. Figure 4 shows qualitative results of this experiment.

Furthermore, we obtained qualitative feedback from facial- and neurosurgeons, who confirmed an applicability of our application for tumor cases (including biopsies) were accurate navigation is not necessary, but a better understanding of three-dimensional relations might still be beneficial. They stated that a 3D visualization provided on a tablet or smartphone screen can easily, quickly and efficiently support clinical diagnosis and treatment pathways in all patients with maxillofacial or cranial tumors, cysts or any other expansive soft or hard tissue processes.

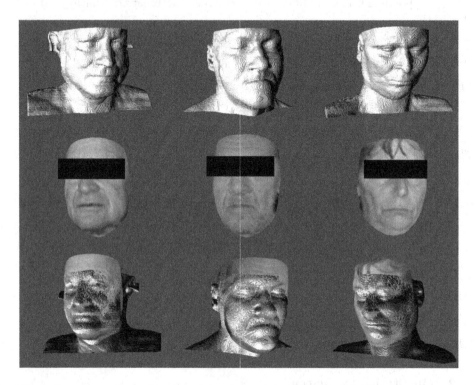

Fig. 3. Reconstruction and registration results of three patients. The first row shows the surface extraction from the patient's CT. The second row shows the 3D reconstructed patient photos (anonymized with a black bar in the eye area due to patient privacy). The bottom row shows the results of the automatic registration between the surface extracted from CT and the 3D photo reconstruction.

4   Discussion

For an objective evaluation of our proposed reconstruction and registration pipeline, we used real patient photos and CT scans from the clinical routine in facial surgery. Since our system requires minimal setup time and expenditure, it can be used for applications where the usage of a commercial navigation

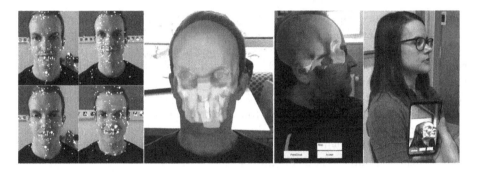

Fig. 4. Examples of several live cases. Left: ARCore is using its SLAM system to generate a point cloud of the environment (white points). Middle: Two examples of our AR visualization on live cases. We display the subject's skulls from the front and from the side. Right: Application usage from the user's perspective, showing the subjects skull and a tumor in the facial area (red). (Color figure online)

Table 1. Total closest point registration error (CPRE), as well as CPRE in x, y and z directions between the 3D reconstructed 2D (patient) photos and the pre-operative CT/MRI scans. For subjects 1–7, CT scans and patient photos were available. Subjects 8–10 are live human subjects with MRI scans. In addition, we present the overall processing time of our system consisting of following steps: (1) sending a 2D photo from smartphone to the server, (2) performing the 2D to 3D reconstruction, (3) registration of the 3D photo reconstruction to the pre-operative CT, and (4) transmission of the registration result back to the mobile application.

Subject		1	2	3	4	5	6	7	8	9	10	Total
CPRE (mm)	Avg	6.1	5.0	5.3	5.7	6.9	5.5	6.0	7.3	6.5	7.7	6.2
	Avg_X	2.5	2.2	2.5	2.7	2.2	2.3	2.2	3.1	3.3	4.0	2.7
	Avg_Y	2.6	3.2	3.2	3.4	5.3	3.4	4.2	4.5	2.7	5.1	3.7
	Avg_Z	3.8	2.1	2.2	2.3	2.1	2.3	2.0	2.8	4.0	3.0	2.4
Time (s)	Mean	10.1	9.5	8.7	10.4	10.4	11.6	10.4	9.5	10.3	11.0	10.2

system, which requires preparation and a setup times of up to 30 min, is not feasible. Our quantitative results show that the VRN is able to reconstruct a patients face sufficiently well for successful registration with volumetric medical imaging data. The demands in the registration accuracy of medical AR systems for image guided interventions are exceptionally high, and conventional mobile hardware is, at this point, usually not able to meet these requirements. Therefore, we focus our system on an application which does not require sub-millimeter precision: Pre-operative visualization, facilitating a rough estimation of the target localization or surgical entry point. Our medical partners attested that for this application, our registration error of around six mm is acceptable. Compared to navigation systems with much higher accuracy (but that might not be available in all medical centers, especially in smaller ones), our application runs on a low-cost smartphone and does not need any preparation or set up time.

In addition, our evaluation grants a deeper insight into VRN. By default, the 3D reconstruction from the 2D photo with the VRN does not provide any real-world size unit, because the distance of the image plane to the face and the person's head size is not known from a single photo. The availability of a millimeter-precise 3D face model from medical scans, such as CT and MRI, gives us the unique opportunity to also estimate the scale of the reconstruction returned by the VRN, which is an aspect that has so far been overlooked by the non-medical computer vision community.

5   Conclusion

In this contribution, we introduced an AR video see-through system for medical applications, which performs image-to-patient registration using only a single patient photo and volumetric imaging data of the patient. Our application runs solely on a mobile device, such as a smartphone or tablet. It allows the visualization of anatomical structures (like bones) and pathological structures (like tumors) for an augmented view of the patient. One main advantage is that our approach, unlike previous work, does not depend on any external devices, such as navigation systems or depth sensors; it only needs a standard smartphone or tablet, which makes it very cost-effective. Furthermore, it works without any markers and does not require complicated calibration.

Our results show that an accurate overlay of virtual content with the real scene can be achieved with our pipeline. Therefore, our system could be used for various medical applications involving the head and face, for example, pre-operative visualization or educational purposes. Future work will evaluate our approach in clinical routine and introduce more sophisticated visualizations and interactions.

Acknowledgment. This work received funding from the Austrian Science Fund (FWF) KLI 678-B31 (enFaced - Virtual and Augmented Reality Training and Navigation Module for 3D-Printed Facial Defect Reconstructions). Further, this work sees the support of CAMed - Clinical additive manufacturing for medical applications (COMET K-Project 871132), which is funded by the Austrian Federal Ministry of Transport, Innovation and Technology (BMVIT), and the Austrian Federal Ministry for Digital and Economic Affairs (BMDW), and the Styrian Business Promotion Agency (SFG), and the TU Graz Lead Project (Mechanics, Modeling and Simulation of Aortic Dissection). Moreover, the Summer Bachelor (SB) Program of the Institute of Computer Graphics and Vision (ICG) of the Graz University of Technology (TU Graz). Finally, we want to point out to our medical online framework Studierfenster (www.studierfenster.at), where an automatic single-shot 3D face reconstruction and registration module has been integrated, and a video tutorial is available on YouTube (3D Face Reconstruction and Registration with Studierfenster: https://www.youtube.com/watch?v=DbbFm9XxlGE).

References

1. Ahn, J., Choi, H., Hong, J., Hong, J.: Tracking accuracy of a stereo-camera-based augmented reality navigation system for orthognathic surgery. J. Oral Maxillofac. Surg. **77**(5), 1070.e1–1070.e11 (2019)
2. Chen, X., et al.: Development of a surgical navigation system based on augmented reality using an optical see-through head-mounted display. J. Biomed. Inform. **55**, 124–131 (2015)
3. Chen, Y., Medioni, G.: Object modelling by registration of multiple range images. Image Vis. Comp. **10**(3), 145–155 (1992)
4. Choi, S., Zhou, Q.Y., Koltun, V.: Robust reconstruction of indoor scenes. In: Conference on Computer Vision and Pattern Recognition (CVPR) (2015)
5. Eggers, G., Mühling, J., Marmulla, R.: Image-to-patient registration techniques in head surgery. Int. J. Oral Maxillofac. Surg. **35**(12), 1081–1095 (2006)
6. Fan, Y., Jiang, D., Wang, M., Song, Z.: A new markerless patient-to-image registration method using a portable 3D scanner. Med. Phys. **41**(10), 101910 (2014)
7. Gsaxner, C., Pepe, A., Wallner, J., Schmalstieg, D., Egger, J.: Markerless image-to-face registration for untethered augmented reality in head and neck surgery. In: Shen, D., et al. (eds.) MICCAI 2019. LNCS, vol. 11768, pp. 236–244. Springer, Cham (2019). https://doi.org/10.1007/978-3-030-32254-0_27
8. Gsaxner, C., Wallner, J., Chen, X., Zemann, W., Egger, J.: Facial model collection for medical augmented reality in oncologic cranio-maxillofacial surgery. Scientific Data **6**(1), 310 (2019)
9. Holz, D., Ichim, A.E., Tombari, F., Rusu, R.B., Behnke, S.: Registration with the point cloud library: a modular framework for aligning in 3-D. IEEE Robot. Autom. Mag. **22**(4), 110–124 (2015)
10. Jackson, A.S., Bulat, A., Argyriou, V., Tzimiropoulos, G.: Large pose 3D face reconstruction from a single image via direct volumetric CNN regression. In: International Conference on Computer Vision (ICCV) (2017)
11. Jayender, J., Xavier, B., King, F., Hosny, A., Black, D., Pieper, S., Tavakkoli, A.: A novel mixed reality navigation system for laparoscopy surgery. In: Frangi, A.F., Schnabel, J.A., Davatzikos, C., Alberola-López, C., Fichtinger, G. (eds.) MICCAI 2018. LNCS, vol. 11073, pp. 72–80. Springer, Cham (2018). https://doi.org/10.1007/978-3-030-00937-3_9
12. Jiang, T., Zhu, M., Chai, G., Li, Q.: Precision of a novel craniofacial surgical navigation system based on augmented reality using an occlusal splint as a registration strategy. Sci. Rep. **9**(1), 501 (2019)
13. Lamecker, H., et al.: Automatic segmentation of mandibles in low-dose CT-data. Int. J. Comput. Assisted Radiol. Surg. **1**, 393 (2006)
14. Lorensen, W.E., Cline, H.E.: Marching cubes: a high resolution 3d surface construction algorithm. ACM Siggraph Comput. Graph. **21**(4), 163–169 (1987)
15. Markelj, P., Tomaževič, D., Likar, B., Pernuš, F.: A review of 3D/2D registration methods for image-guided interventions. Med. Image Anal. **16**(3), 642–661 (2012)
16. Maruyama, K., et al.: Smart glasses for neurosurgical navigation by augmented reality. Operative Neurosurgery **15**(5), 551–556 (2018)
17. McCann, M.T., Nilchian, M., Stampanoni, M., Unser, M.: Fast 3d reconstruction method for differential phase contrast x-ray CT. Optics Express **24**(13), 14564–14581 (2016)
18. Meulstee, J.W., et al.: Toward holographic-guided surgery. Surgical Innov. **26**(1), 86–94 (2019)

19. Olabarriaga, S.D., Smeulders, A.W.: Interaction in the segmentation of medical images: a survey. Med. Image Anal. **5**(2), 127–142 (2001)
20. Orentlicher, G., Goldsmith, D., Horowitz, A.: Applications of 3-dimensional virtual computerized tomography technology in oral and maxillofacial surgery: current therapy. J. Oral Maxillofacial Surgery **68**(8), 1933–1959 (2010)
21. Pepe, A., et al.: Pattern recognition and mixed reality for computer-aided maxillofacial surgery and oncological assessment. In: Proceedings Biomedical Engineering International Conference (BMEiCON), pp. 1–5. IEEE, January 2019
22. Pepe, A., et al.: A marker-less registration approach for mixed reality–aided maxillofacial surgery: a pilot evaluation. J. Dig. Imag. **32**(6), 1008–1018 (2019). https://doi.org/10.1007/s10278-019-00272-6
23. Rusu, R.B., Blodow, N., Beetz, M.: Fast point feature histograms (FPFH) for 3D registration. In: ICRA (2009)
24. Sielhorst, T., Feuerstein, M., Navab, N.: Advanced medical displays: a literature review of augmented reality. J. Disp. Technol. **4**(4), 26 (2008)
25. Sylos Labini, M., Gsaxner, C., Pepe, A., Wallner, J., Egger, J., Bevilacqua, V.: Depth-awareness in a system for mixed-reality aided surgical procedures. In: Huang, D.-S., Huang, Z.-K., Hussain, A. (eds.) ICIC 2019. LNCS (LNAI), vol. 11645, pp. 716–726. Springer, Cham (2019). https://doi.org/10.1007/978-3-030-26766-7_65
26. Tucker, S., et al.: Comparison of actual surgical outcomes and 3-dimensional surgical simulations. J. Oral Maxillofacial Surg. **68**(10), 2412–2421 (2010)
27. Wallner, J., et al.: Clinical evaluation of semi-automatic open-source algorithmic software segmentation of the mandibular bone: practical feasibility and assessment of a new course of action. PLoS ONE **13**(5), 156–165 (2018)
28. Wallner, J., Schwaiger, M., Hochegger, K., Gsaxner, C., Zemann, W., Egger, J.: A review on multiplatform evaluations of semi-automatic open-source based image segmentation for cranio-maxillofacial surgery. In: Computer Methods and Programs in Biomedicine, p. 105102 (2019)
29. Wang, J., Shen, Yu., Yang, S.: A practical marker-less image registration method for augmented reality oral and maxillofacial surgery. Int. J. Comput. Assisted Radiol. Surg. **14**(5), 763–773 (2019). https://doi.org/10.1007/s11548-019-01921-5
30. Weber, M., Wild, D., Wallner, J., Egger, J.: A client/server based online environment for the calculation of medical segmentation scores. In: EMBC, pp. 3463–3467 (2019). https://doi.org/10.1109/EMBC.2019.8856481
31. Wild, D., Weber, M., Wallner, J., Egger, J.: Client/server based online environment for manual segmentation of medical images. CoRR abs/1904.08610 (2019). http://arxiv.org/abs/1904.08610

A Baseline Approach for AutoImplant: The MICCAI 2020 Cranial Implant Design Challenge

Jianning Li[1,2]($\boxtimes$) , Antonio Pepe[1,2] , Christina Gsaxner[1,2,3] ,
Gord von Campe[4] , and Jan Egger[1,2,3]

[1] Institute of Computer Graphics and Vision, Graz University of Technology,
Graz, Austria
{jianning.li,egger}@icg.tugraz.at
[2] Computer Algorithms for Medicine Laboratory (Café-Lab), Graz, Austria
[3] Department of Oral and Maxillofacial Surgery, Medical University of Graz,
Graz, Austria
[4] Department of Neurosurgery, Medical University of Graz, Graz, Austria

Abstract. In this study, we present a baseline approach for AutoImplant (https://autoimplant.grand-challenge.org/) – the cranial implant design challenge, which can be formulated as a volumetric shape learning task. In this task, the defective skull, the complete skull and the cranial implant are represented as binary voxel grids. To accomplish this task, the implant can be either reconstructed directly from the defective skull or obtained by taking the difference between a defective skull and a complete skull. In the latter case, a complete skull has to be reconstructed given a defective skull, which defines a volumetric shape completion problem. Our baseline approach for this task is based on the former formulation, i.e., a deep neural network is trained to predict the implants directly from the defective skulls. The approach generates high-quality implants in two steps: First, an encoder-decoder network learns a coarse representation of the implant from downsampled, defective skulls; The coarse implant is only used to generate the bounding box of the defected region in the original high-resolution skull. Second, another encoder-decoder network is trained to generate a fine implant from the bounded area. On the test set, the proposed approach achieves an average dice similarity score (DSC) of 0.8555 and Hausdorff distance (HD) of 5.1825 mm. The codes are available at https://github.com/Jianningli/autoimplant.

Keywords: Shape learning · Cranial implant design · Cranioplasty · Deep learning · Skull reconstruction · Volumetric shape completion

1 Introduction

In current clinical practice, the process of cranial implant design and manufacturing is performed externally by a third-party supplier. The process usually

---

https://autoimplant.grand-challenge.org/.

© Springer Nature Switzerland AG 2020
T. Syeda-Mahmood et al. (Eds.): ML-CDS 2020/CLIP 2020, LNCS 12445, pp. 75–84, 2020.
https://doi.org/10.1007/978-3-030-60946-7_8

involves costly commercial software and highly-trained professional users [1]. A fully automatic, low-cost and in-Operation Room (in-OR) design and manufacturing of cranial implants can bring significant benefits and improvements to the current clinical workflow for cranioplasty [11]. Previous work has seen the development of freely available CAD tools for cranial implant design [3,5,7,12], whereas these approaches are still time-consuming and require human interaction. These approaches tend to exploit the geometric information of skull shape. For example, by finding the symmetry plane of the skull and fill the defected region by mirroring [2]. Considering that human skulls are not strictly symmetric, mirroring is not an optimal solution.

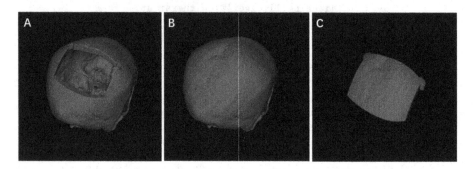

Fig. 1. Illustration of a defective skull (A), complete skull (B) and the implant (C). The defective skull is created by removing a bony part (i.e., the implant) from the complete skull.

The AutoImplant Challenge [6] aims at solving the problem of automatic cranial implant design in a data-driven manner, without relying explicitly on geometric shape priors of human skulls. As suggested by the organizers, cranial implant design can be formulated as a volumetric shape learning task, where the shape of the implant can be learnt directly or indirectly from the shape of a defective skull [11]. On the one hand, the shape of the implant can be directly learnt from a defective skull. On the other hand, by learning to fill the defected region on a defective skull, a completed skull can be produced. The implant can then be obtained indirectly by taking the difference between the completed skull and the defective skull. In this sense, cranial implant design is being formulated as a shape completion problem [4,8,14–16]. A relevant study was conducted by Morais et al, where an encoder-decoder network is used to predict a complete skull from a defective skull [13]. However, the study deals with very coarse skulls of low dimensionality (30^3, 60^3 and 120^3) extracted from MRI data, whereas in practice, the common imaging modality used for head scan acquisition is computed tomography (CT), with a typical resolution of $512 \times 512 \times Z$. [9,10] further extended shape completion to high-resolution volumetric CT skulls by using a patch-based training strategy. In this study, we primarily elaborate on the former formulation, i.e., given a defective skull, we directly predict the shape of the implant, which is a challenging task as the implant has to be congruent with the defective skull in terms of shape, bone thickness and boundaries of the defected region [11].

The defective skulls in the challenge datasets are created artificially out of complete skulls. By doing so, we have a ground truth for supervised training for either of the two above mentioned problem formulations. For direct implant generation, the ground truth is the implant, which is the region removed from a complete skull. For skull shape completion, the ground truth is the original complete skull. The input of either formulation is the defective skull. Real surgical defects from a craniotomy surgery are usually more complex and irregular than the artificial defects. However, we expect that the deep learning networks trained on artificial defects can be generalized to the real surgical defects in craniotomy, which requires that the networks should be robust as to the shape, position and size of the defects.

2 Dataset

This section briefly introduces the challenge dataset used by this baseline approach. The 200 unique skulls (100 for training and 100 for testing) are selected from CQ500 (http://headctstudy.qure.ai/dataset), which is a public collection and contains 491 anonymized head CT scans in DICOM format. Considering that the datasets are acquired from patients with various head pathologies, we discarded the scans that present a severe skull deformity or damage. Lower-quality scans (e.g., z-spacing above 1 mm) were also discarded. The dimension of these skulls is $512 \times 512 \times Z$, where Z is the number of axial slices. For ease of use, the selected DICOM scans were converted to the NRRD format. To extract the binary skull data, a fixed threshold (Hounsfield units values from 150 to maximum) was applied to the CT scans. As the thresholding also preserves the CT table head holder, which has a similar density to the bony structures, we used 3D connected component analysis to automatically remove this undesired component. The last step is to generate an artificial surgical defect on each skull, which was accomplished by removing a bony structure from the skull. Our github repository (https://github.com/Jianningli/autoimplant) contains codes for generating the synthetic defects in an automated manner, which also allows user-specified size and location of the defect to be created. The data processing step is summarized as follows:

1. **DICOM Selection:** 200 High quality DICOM files were selected.
2. **NRRD Conversion:** DICOM files were converted into NRRD format.
3. **Skull Extraction:** Skulls were extracted using thresholding (150 HU-Max).
4. **CT Table Removal:** The CT table head holders were removed.
5. **Hole Injection:** On each skull, an artificial surgical defect was injected.

Figure 1 shows a defective skull, the corresponding original skull and the implant (i.e., the removed part) in the training set. The skull defects shown in Fig. 1 (A) are representative of those of the 100 training datasets and 100 test datasets.

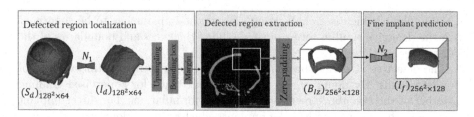

Fig. 2. The workflow of the proposed approach. Theoretically, $\mathbf{N}_1$ and $\mathbf{N}_2$ can be any network with an encoder-decoder architecture. In our study, the configurations of $\mathbf{N}_1$ and $\mathbf{N}_2$ are shown in Fig. 3.

3   Method

The proposed implant generation scheme is illustrated in Fig. 2 and consists of three steps. **First**, an encoder-decoder network $\mathbf{N}_1$ learns to infer a coarse implant $(\mathbf{I})_{128^2 \times 64}$ directly from a coarse defective skull $(\mathbf{S}_d)_{128^2 \times 64}$, which is downsampled from the original high-resolution defective skull $(\mathbf{S}_d)_{512^2 \times Z}$. This allows to contain the requirements of GPU memory:

$$(\mathbf{I}_c)_{128^2 \times 64} \overset{\mathbf{N}_1}{\leftarrow} (\mathbf{S}_d)_{128^2 \times 64}. \tag{1}$$

Second, we calculate the bounding box $(\mathbf{B}_I)_{X_B \times Y_B \times 128}$ of the coarse implant predicted by $\mathbf{N}_1$, which is then used to localize the defected region on the high-resolution defective skull. X_B and Y_B are dimensions of the bounding box in x/y volume axis. In the z axis, we fix the dimension to 128 (the maximum z dimension of the defected area in the challenge dataset is smaller than 128). Considering that the bounding box tightly encloses the defected region in the x/y axis, a margin $\mathbf{m}$ is used to keep some surrounding information around the defected region. In order to get a fixed bounding box dimension $(\mathbf{B}_{Iz})_{256^2 \times 128}$, zero-padding is applied. **Third**, a second encoder-decoder network $\mathbf{N}_2$ learns to infer the fine implants $(\mathbf{I}_f)_{256^2 \times 128}$ from the bounded region of the high-resolution defective skulls:

$$(\mathbf{I}_f)_{256^2 \times 128} \overset{\mathbf{N}_2}{\leftarrow} (\mathbf{B}_{Iz})_{256^2 \times 128} \tag{2}$$

The detailed architecture of $\mathbf{N}_1$ and $\mathbf{N}_2$ is shown in Fig. 3. As the input size of $\mathbf{N}_2$ is larger than that of $\mathbf{N}_1$, the complexity of $\mathbf{N}_2$ has to be significantly reduced compared to $\mathbf{N}_1$ in order to get the network running on the limited GPU memory. In particular, the kernel size of all convolutional layers in $\mathbf{N}_1$ is five, whereas the kernel size for $\mathbf{N}_2$ is only three. The number of feature maps of each layer for $\mathbf{N}_2$ is also significantly reduced, resulting in a total of 0.6538 million trainable parameters, compared to 82.0766 million parameters for $\mathbf{N}_1$. Figure 2 shows the *input/output* of $\mathbf{N}_1$ and $\mathbf{N}_2$. $\mathbf{N}_1$ takes as input a downsampled defective skull and produces a coarse implant prediction. $\mathbf{N}_2$ takes as input a zero-padded version of the high-resolution defected area delimited by the bounding box $(\mathbf{B}_{Iz})_{256^2 \times 128}$ and produces a prediction of the corresponding fine implant.

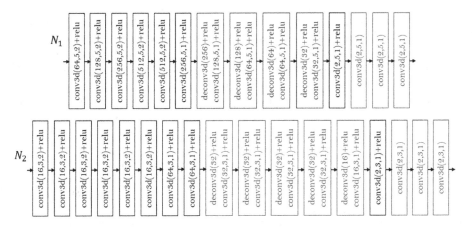

Fig. 3. Detailed configurations of $\mathbf{N}_1$ and $\mathbf{N}_2$. The parameters in conv3d denote the number of feature maps, kernel size and strides respectively. The parameter in deconv3d denotes the number of feature maps. For all the deconvolutional layers (deconv3d), the kernel size and strides are set to 4 and 2 respectively.

The bounding box and the amount of zero-padding are calculated as follows:

Bounding Box. The bounding box of an implant is calculated by finding the coordinates of the first and last non-zero values in the projected x/y plane of the image volume $(\mathbf{V})_{512^2 \times Z}$. The predicted coarse implant $(\mathbf{I})_{128^2 \times 64}$ is first upsampled to its original dimension $512^2 \times Z$ using a (order two) spline interpolation before the bounding box is calculated. The bounding box $(\mathbf{B}_I)_{X_B \times Y_B \times 128}$ tightly encloses the defected region in the original high-resolution skull. We apply an additional margin $\mathbf{m} = 20$ in the x/y direction to also enclose a portion of the surrounding skull, which facilitates the learning task. The x/y dimension of the resulting bounding box becomes: $X_B + 2m$ and $Y_B + 2m$.

Zero-Padding. As the dimension of each bounding box is different, we apply zero-padding on the bounding boxes to obtain inputs with a fixed dimension $(\mathbf{B}_{Iz})_{256^2 \times 128}$ for the deep neural network $\mathbf{N}_2$. Zero-padding is done by moving the bounding box to the middle of an all-zero volume of dimension $256^2 \times 128$.

4 Experiments and Results

$\mathbf{N}_1$ and $\mathbf{N}_2$ were consecutively trained on a machine equipped with one GPU NVIDIA GeForce GTX 1070 Ti, which presents a limited GPU memory of 8 GB. First, $\mathbf{N}_1$ was trained on downsampled defective skulls. Once the training of $\mathbf{N}_1$ was completed, we used $\mathbf{N}_1$ to produce coarse implants on the training set. Then, the coarse implants were upsampled to their original size of each corresponding training sample. Second, the upsampled implants were used to calculate the bounding box of the defected region on the high-resolution defective skulls. The bounding box, extended by a margin of $2 \times m$ to include a portion of skull, was used to train $\mathbf{N_2}$. The networks were trained on the 100 data pairs provided by the AutoImplant challenge, without using any additional dataset or defect shapes for data augmentation. Additionally, it needs to be considered that the performance of $\mathbf{N_2}$ depends on the accuracy, or failure rate, of $\mathbf{N_1}$. In both cases the batch size was set to one. We employed dice loss as a loss function, which measures the shape similarity between a predicted implant and its corresponding ground truth implant. Figure 4 shows the *step/loss* curve during training. Shape similarity between the predicted implant and the ground truth is quantitatively evaluated using the Dice similarity score (DSC), the symmetric Hausdorff distance (HD) and the reconstruction error (RE). The RE for each test case is defined as the false voxel prediction rate as in [13]:

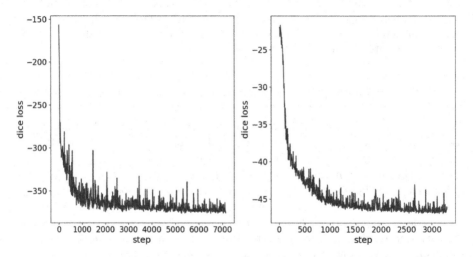

Fig. 4. The *step/loss* curve of $\mathbf{N_1}$ (left) and $\mathbf{N_2}$ (right) during training. Both $\mathbf{N_1}$ and $\mathbf{N_2}$ are trained using Adam optimizer with a fixed learning rate of 0.0001. The dice loss function is implemented as the negative of dices score. We multiply a large number (in our case $200M$) with the loss function as we notice that the dice loss function alone tends to produce very small values, which hinders the learning process.

$$RE = \frac{\sum |\mathbf{P} - \mathbf{G}|}{N} \qquad (3)$$

$(\mathbf{P})_{512^2 \times Z}$ and $(\mathbf{G})_{512^2 \times Z}$ are the fine implant produced by $\mathbf{N_2}$ and its corresponding ground truth, respectively. $\sum |\mathbf{P} - \mathbf{G}|$ represents the total number of voxels in $\mathbf{P}$ that are different from $\mathbf{G}$. $N = 512^2 \times Z$ is the total number of voxels in the volume. Note that, in order to calculate the metrics – DSC, HD, and RE – against the ground truth, which has a $512^2 \times Z$ dimension, the corresponding inverse process of zero-padding and bounding box was applied to

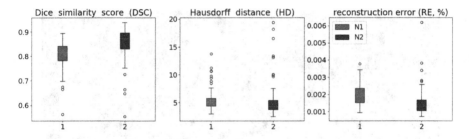

Fig. 5. Boxplots of DSC (left), HD (middle) and RE (right) of the 100 test cases.

the prediction $(\mathbf{I}_f)_{256^2 \times 128}$ from $\mathbf{N}_2$ so that the prediction $(\mathbf{P})_{512^2 \times Z}$ was of the same dimension as that of the ground truth $(\mathbf{G})_{512^2 \times Z}$. Similarly, to calculate the metrics for $\mathbf{N}_1$, the coarse implants $(\mathbf{I}_c)_{128^2 \times 64}$ were upsampled to their corresponding original dimensions $512^2 \times Z$ using interpolation. To provide the HD in millimeters (mm), we considered the actual image spacing of each test case, which is provided in the header of the NRRD files. Table 1 shows the mean value of DSC, HD, and RE on the 100 test cases. The corresponding boxplot is shown in Fig. 5.

Table 1. Quantitative Results

	DSC	HD (mm)	RE (%)
$\mathbf{N}_1$	0.8097	5.4404	0.20
$\mathbf{N}_2$	0.8555	5.1825	0.15

Figure 6 gives an illustration of the automatic implant generation results in 3D for five test cases (A-E). We can see that the implants from $\mathbf{N}_1$ are coarse (second column), lacking geometric details compared to the ground truth (fourth column). The reason is that the implants are learnt from downsampled skulls $((\mathbf{S}_d)_{128^2 \times 64})$, which are already deviating from the original high-resolution skulls. In comparison, $\mathbf{N}_2$ produces fine, high-quality implants (third column), which are close to the ground truth, as $\mathbf{N}_2$ learns directly from high-resolution skull shapes. We also empirically noticed how $\mathbf{N}_2$ captures highly intricate details such as the smoothness of the implant surface and the details of the small roundish corners of the implants, which are not well preserved in the coarse implants generated by $\mathbf{N}_1$. Furthermore, (A'-E') show how the fine implants generated by $\mathbf{N}_2$ match with the defected region on the defective skulls in terms of shape and bone thickness. Figure 7 (A) shows a zooming in of the coarse implant (left) generated by $\mathbf{N}_1$, the fine implant (middle) generated by $\mathbf{N}_2$ and the ground truth (right). Figure 7 (B) show how the shape of the predicted implant (red) matches with that of the ground truth (white) in 2D axial, sagittal and coronal view.

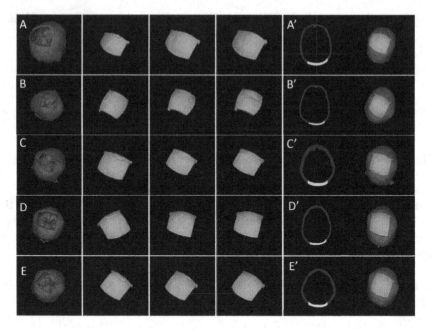

Fig. 6. (A)–(E) implant prediction results on five evaluation cases. From left to right: the input defective skulls; the coarse implants from N_1; the fine implant predictions from N_2; the ground truth. (A')–(E') overlay of the implants from N_2 on the defective skulls in 2D axial view (fifth column) and in 3D (sixth column). To differentiate them, different colors are used for the implants (gray) and skulls (red). (Color figure online)

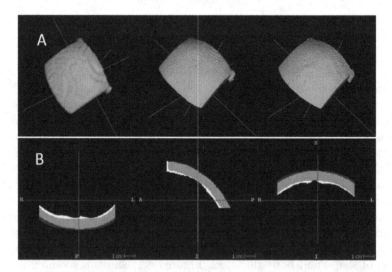

Fig. 7. (A): A zooming in of the coarse implant (left), fine implant (middle) and the ground truth (right). (B): how the shape of the predicted implant (red) matches that of the ground truth (white) in 2D axial, sagittal and coronal view. (Color figure online)

5   Conclusion and Future Improvement

The contribution of this study is threefold. **First**, we demonstrated that a fully data-driven approach without using geometric priors can be effective in high-resolution volumetric shape learning. An encoder-decoder network (N_1) can directly learn to reconstruct the missing part from a defective skull. **Second**, we show that an encoder-decoder network (N_2) does not need to see the entire skull shape to predict the missing part. Instead, the learning can be based only on the defected region with limited surrounding shape information. **Third**, we propose a **coarse-to-fine** framework for high-quality implants generation. Under the framework, the coarse implants are first inferred from coarse, downsampled skulls. Then, the high-quality implants can be inferred from the defected regions of high-resolution skulls. The technique is especially useful when only limited GPU memory is available whereas the dimension of the data to be processed is high (e.g., in our case, $512 \times 512 \times Z$). The study can serve as a baseline for the AutoImplant challenge. N_2 and N_1 can be further designed to improve the respective performance. Currently, the experiments are only carried out on skulls with synthetic defects. Our future study will be focused on the generation of more realistic cranial implants, which are more irregular and complex than the synthetic ones used in this study, so that a fully automated cranial implant design system can be implemented in clinical practice.

Acknowledgment. This work received the support of CAMed - Clinical additive manufacturing for medical applications (COMET K-Project 871132), which is funded by the Austrian Federal Ministry of Transport, Innovation and Technology (BMVIT), and the Austrian Federal Ministry for Digital and Economic Affairs (BMDW), and the Styrian Business Promotion Agency (SFG). Further, this work received funding from the Austrian Science Fund (FWF) KLI 678-B31 (enFaced - Virtual and Augmented Reality Training and Navigation Module for 3D-Printed Facial Defect Reconstructions) and the TU Graz Lead Project (Mechanics, Modeling and Simulation of Aortic Dissection). Moreover, we want to point out to our medical online framework Studierfenster (www.studierfenster.at) [17], where an automatic cranial implant design system has been incorporated. Finally, we thank the creators of the CQ500 dataset (http://headctstudy.qure.ai/dataset).

References

1. Digital evolution of cranial surgery. A case study by Renishaw PLC in New Mills, Wotton-under-Edge Gloucestershire, GL12 8JR United Kingdom (2017)
2. Angelo, L., Di Stefano, P., Governi, L., Marzola, A., Volpe, Y.: A robust and automatic method for the best symmetry plane detection of craniofacial skeletons. Symmetry **11**, 245 (2019). https://doi.org/10.3390/sym11020245
3. Chen, X., Xu, L., Li, X., Egger, J.: Computer-aided implant design for the restoration of cranial defects. Sci. Rep. **7**, 1–10 (2017). https://doi.org/10.1038/s41598-017-04454-6
4. Dai, A., Qi, C.R., Nießner, M.: Shape completion using 3D-encoder-predictor CNNS and shape synthesis. In: 2017 IEEE Conference on Computer Vision and Pattern Recognition (CVPR), pp. 6545–6554 (2016)

5. Egger, J.: Interactive reconstructions of cranial 3D implants under MeVisLab as an alternative to commercial planning software. PLoS ONE **12**, 20 (2017). https://doi.org/10.1371/journal.pone.0172694

6. Egger, J., et al.: Towards the automatization of cranial implant design in cranioplasty (2020). https://doi.org/10.5281/zenodo.3715953

7. Gall, M., Li, X., Chen, X., Schmalstieg, D., Egger, J.: Computer-aided planning and reconstruction of cranial 3d implants. In: EMBC, pp. 1179–1183 (2016). https://doi.org/10.1109/EMBC.2016.7590915

8. Han, X., Li, Z., Huang, H., Kalogerakis, E., Yu, Y.: High-resolution shape completion using deep neural networks for global structure and local geometry inference. In: 2017 IEEE International Conference on Computer Vision (ICCV), pp. 85–93 (2017)

9. Li, J.: Deep learning for cranial defect reconstruction. Master's thesis, Graz University of Technology, January 2020

10. Li, J., Egger, J.: Towards the automatization of cranial implant design for 3D printing. ResearchGate (2019). https://doi.org/10.13140/RG.2.2.16144.56324

11. Li, J., Pepe, A., Gsaxner, C., Egger, J.: An online platform for automatic skull defect restoration and cranial implant design. arXiv:2006.00980 (2020)

12. Marzola, A., Governi, L., Genitori, L., Mussa, F., Volpe, Y., Furferi, R.: A semi-automatic hybrid approach for defective skulls reconstruction. Comput. Aided Des. Appl. **17**, 190–204 (2019). https://doi.org/10.14733/cadaps.2020.190-204

13. Morais, A., Egger, J., Alves, V.: Automated computer-aided design of cranial implants using a deep volumetric convolutional denoising autoencoder. In: Rocha, Á., Adeli, H., Reis, L.P., Costanzo, S. (eds.) WorldCIST'19 2019. AISC, vol. 932, pp. 151–160. Springer, Cham (2019). https://doi.org/10.1007/978-3-030-16187-3_15

14. Sarmad, M., Lee, H.J., Kim, Y.M.: RL-GAN-Net: A reinforcement learning agent controlled GAN network for real-time point cloud shape completion. In: 2019 IEEE/CVF Conference on Computer Vision and Pattern Recognition (CVPR), pp. 5891–5900 (2019)

15. Stutz, D., Geiger, A.: Learning 3D shape completion under weak supervision. Int. J. Comput. Vis., **15**, 1–20 (2018)

16. Sung, M., Kim, V.G., Angst, R., Guibas, L.J.: Data-driven structural priors for shape completion. ACM Trans. Graph. **34**, 175:1–175:11 (2015)

17. Weber, M., Wild, D., Wallner, J., Egger, J.: A client/server-based online environment for the calculation of medical segmentation scores. In: EMBC, pp. 3463–3467, July 2019. https://doi.org/10.1109/EMBC.2019.8856481

Adversarial Prediction of Radiotherapy Treatment Machine Parameters

Lyndon Hibbard[(✉)]

Elekta Inc., St. Charles, MO 63303, USA
Lyn.Hibbard@elekta.com
http://www.elekta.com/

Abstract. Modern external-beam cancer radiotherapy applies prescribed radiation doses to tumor targets while minimally affecting nearby
vulnerable organs-at-risk (OARs). Creating a clinical plan is difficult and
time-consuming with no guarantee of optimality. Knowledge-based planning (KBP) mitigates this uncertainty by guiding planning with probabilistic models based on populations of prior clinical-quality plans. We have
developed a KBP-inspired planning model that predicts plans as realizations of the treatment machine parameters. These are tuples of linear accelerator (Linac) gantry angles, multi-leaf collimator (MLC) apertures that
shape the beam, and aperture-intensity weights that can be represented
graphically in a coordinate frame isomorphic with projections (beam's-eye
views) of the patient's target anatomy. These paired data train CycleGAN
networks that estimate the MLC apertures and weights for a novel patient,
thereby predicting a treatment plan. The dosimetric properties of the predicted plans agree with clinical plans to within a few percent, and take far
less time to compute. The predicted plans can serve as lower bounds on plan
quality, and as initializations for MLC aperture shape and weight refinement.

Keywords: Radiotherapy · Deep learning · Plan prediction

1 Introduction

Cancer radiotherapy is based on two main premises. The first is that tumor cells
are less competent to repair DNA strand breaks due to radiation than nearby
normal cells. The second is that this biological difference is best exploited by irra-
diating the tumor with (possibly many) shaped beams of radiation performing
a kind of forward tomography, matching the 3D target shape with a 3D dose.
Planning a treatment personalized for each patient requires the physician to
balance tumor dose with inevitable damage to nearby vulnerable organs-at-risk
(OARs). Planning difficulty is increased by anatomy complexity and by a number of dose constraints to individual targets and OARs that may be greater than
the number of structures. Maximizing plan quality by adjusting plan parameters

Supported by Elekta, Inc.

T. Syeda-Mahmood et al. (Eds.): ML-CDS 2020/CLIP 2020, LNCS 12445, pp. 85–94, 2020.
https://doi.org/10.1007/978-3-030-60946-7_9

is difficult as the effects of any change can be difficult to anticipate. Treatment planning programs (TPPs) provide accurate models for dose physics, but they do not provide assurance that any given plan is close to the best possible, or provide direction to produce a better plan.

To mitigate planning uncertainty, knowledge-based planning (KBP) models dose properties [1–4] and dose distributions [5–8] by sampling from probabilistic models learned from populations of clinical plans. The predicted properties must then be converted into linear accelerator (Linac)/multileaf collimator (MLC) parameters that initialize a TPP dose calculation.

Here we report a new KBP-inspired approach that learns to predict the Linac/MLC parameters directly, bypassing intermediate steps to produce realizations of accurate plans.

2 Methods

2.1 Patient Data Preparation and Treatment Planning

This study used 178 prostate datasets, planned identically, each starting with planning CT images and anatomy contours, to obtain volume modulated arc therapy (VMAT) plans [9]. The cases were anonymized and the contours were curated to conform to the RTOG 0815 standard [10]. The anatomy structures included the prostate, bladder, rectum, femoral heads, anus, penile bulb, seminal vesicles and the patient external contour. The targets are a pair of nested planning target volumes (PTVs); PTV1 enclosed the prostate plus a margin and PTV2 enclosed PTV1 and the seminal vesicles plus an additional margin. We obtained optimal dose fluence maps using the Erasmus-iCycle program [11,12] following the objectives and constraints in [13]: PTV1 MAX = 104% of 78 Gy, PTV2 (outside PTV1) MAX = 104% of 72.2 Gy. The fluence maps were input to Elekta Monaco (v. 5.11) to obtain clinical, deliverable VMAT plans, or the ground truth (GT) plans. The 178 cases were divided into training (138), validation (20), and test (20) sets.

The VMAT treatment is specified by the apertures' shapes and their intensity weights at 100 or more discrete linac gantry angles. The MLC creates a 2D beam profile, normal to the beam direction, by opening gaps between opposing banks of tungsten leaves. The gantry angles, aperture leaf positions, and intensity weights are collectively the control points. During treatment the linac gantry revolves continuously around the patient and the instantaneous linac radiation output and MLC beam shapes at any real angle are interpolated from the flanking control points. In this way, the control points or machine parameters constitute the deliverable treatment plan.

2.2 Data Reformatting for Supervised Learning

The target anatomy and the VMAT control points have fundamentally different representations–anatomies are rectilinear arrays of intensities and control

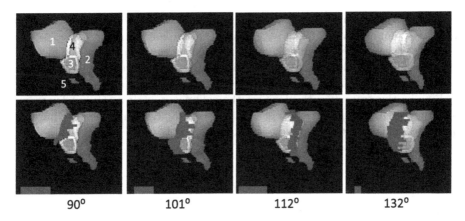

90°   101°   112°   132°

Fig. 1. Anatomy projections (top) and overlayed ground truth apertures (red, bottom) for a prostate target at four gantry angles. The challenge for radiotherapy is to irradiate the target volume (prostate (3) and seminal vesicles (4) plus margins) with the prescribed dose while sparing the adjacent organs-at-risk (OARs; bladder (1), rectum (2), and penile bulb (5)). The red bar length encodes the apertures' weights. The combined gantry angles, apertures and aperture weights are sufficient to specify a treatment plan. The apertures typically sweep back and forth across the target several times during one 2π arc to produce a clinical dose distribution. See text for more details. (Color figure online)

points are vectors of real MLC leaf positions, plus scalar aperture weights and gantry angles. We connect these data domains by matching graphical images of the MLC leaf apertures with beam's-eye-view projection images of the corresponding anatomy taken at the same gantry angle, and projected onto the normal plane containing the treatment isocenter (Fig. 1). That isocenter that is the point on the gantry axis of rotation that is also the perpendicular projection of the center of the MLC. The isocenter is also a reference point in the target volume providing a common origin for both the anatomy and treatment machine coordinate frames.

The CT images and the structure contours were transformed into OAR and PTV 3D binary masks, weighted, summed, and reformatted to a single 3D 8-bit image in the original CT frame. They were designed to emphasize the target volumes' edges from all directions. These 3D anatomy maps were resampled as projections using the forward projection function of the cone beam CT reconstruction program RTK [14]. Graphical images of the apertures were created from the MLC leaf positions and aligned with the plan isocenter and oriented and scaled to match the projections. The projection and aperture images were created at 128 equi-spaced gantry angles producing one projection and one aperture image at each angle. Each of these 2D images has dimensions 128×128 pixels (1.5 mm pixel spacing throughout) so the complete data for each patient consisted of two 128^3 volumes–one a stack of the projections and the second a stack of the aperture graphic images. An example of this data is shown in Fig. 1.

The graphical images also include a bar at the lower left whose length encodes the weight (cumulative meterset weight) or dose increment for that control point. For 100 or more control points, the weights typically range in value from 0.001 to 0.01 and the space alloted for the bar length easily enables weight representation to within about 1% of a typical weight increment.

Finally, the ground truth plans set the collimator angle to 5° but for training the aperture graphic was depicted at 0°. After training, the test inferred apertures' leaf positions were re-written back into DICOM RT Plan files with collimator settings at 5°.

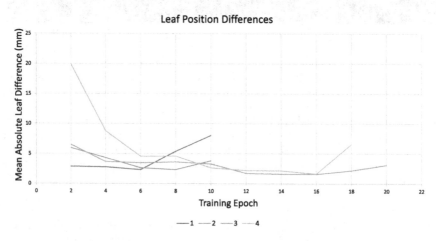

Fig. 2. Training histories for 3D CycleGAN networks learning MLC apertures and weights. Plotted here are the mean absolute MLC leaf position differences in millimeters (MALD) between GT and CycleGAN plans for the 20 test cases not used in training. The four histories correspond to four training regimes differing by learning rate schedule. A network model was saved at every 2 epochs to generate test plans and leaf position differences. All the training regimes produced minima at or below 2.3 mm. (See Table 1.)

2.3 Deep Learning Network and Training

Generative adversarial networks (GANs) have recently become essential technologies for data modeling and inference [15,16]. Here we use a GAN to perform voxel-wise binary classification of aperture and weight-bar voxels. That is, is a voxel in an aperture or weight-bar or not? (Fig. 1).

The results presented here were obtained with a CycleGAN network [17] but with paired data conditioned as for the pix2pix conditional GAN [18]. The full objective function is that used in [17] plus the identity loss. The discriminator network is a (70, 70, 70) PatchGAN with layer structure C18-C36-C76-C152-C304-C304 ([17], Appendix) where C is a convolution/instance normalization [19]/LeakyReLU block with kernel (4, 4, 4) and stride (2, 2, 2), and k equals the number of filters for each Ck. The last layer is a convolution to produce a 1D output with mean squared error loss. The generator is a 3D U-Net [20,21] with six levels each for the encoding and decoding branches. The encoding branch layers are E18-E36-E54-E72-E90-E108 where each Ek contains two convolution/LeakyReLU/batch normalization [22] blocks with a densenet array of skip connections, kernel (3, 3, 3) and stride (1, 1, 1), and k equals the number of filters. The decoding branch is likewise D108-D90-D72-D54-D36-D18 where each block Dk contains a convolution transpose layer and two convolution/LeakyReLU/batch normalization blocks with kernel and stride dimensions and densenet skip connections like those of the encoding branch. The training batch size was a single 3D volume pair. The four CycleGAN losses were weighted as: adversarial (F), adversarial (G), cycle consistent, identity = 10:10:5:1. The numbers of filters used were the largest possible given the available gpu memory. The ADAM optimizer [23] was used throughout. The network was coded in Keras [24] with TensorFlow [25] and based partly on a Keras example CycleGAN code [26]. Four studies are presented here with the training conditions listed in Table 1.

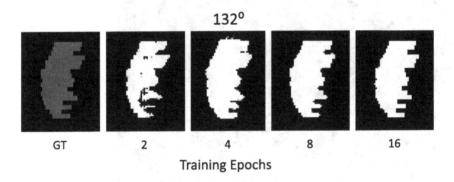

Fig. 3. Demonstration of the visual agreement of the inferred apertures with the corresponding GT aperture. This is the 132° aperture from Fig. 1, from Study 4, Epoch 16 in Fig. 2. The inferred leaf positions are extracted from left and right side boundaries of the white figures, and those at Epoch 16 agree most closely with the ground truth aperture.

The training goal is a model whose inferred apertures and weights lead to an accurate dose distribution, but this calculation is too expensive for all the possbile epoch-interval models. Since MLC leaf positions (encoded by the graphic's left and right edges) dominate the weights in the TPP dose objective function, we computed the mean absolute leaf position differences (MALD) between the CycleGAN and GT plans as a surrogate for dose distribution quality. Figure 2 plots the MALD values for four learning studies where models were saved at every two epochs. Each of these studies produced at least one model with a MALD of 2.3 mm or less. Table 1 lists the minimum-MALD epoch and MALD

Table 1. Learning rate (LR) studies.

Study	Generator LR	Discriminator LR	LR schedule	min-Epoch	MALD
1	10^{-3}	10^{-6}	LRs constant all epochs	6	2.31
2	10^{-4}	10^{-4}	LRs constant all epochs	8	2.32
3	10^{-4}	10^{-4}	LRs/4.0 at Epoch 5	14	1.62
4	10^{-3}	10^{-4}	LRs constant all epochs	16	1.65

value for the four studies. Figure 3 shows the gradual improvement of the 132°
aperture from Study 4 (Fig. 2) at the epochs indicated.

3   Results

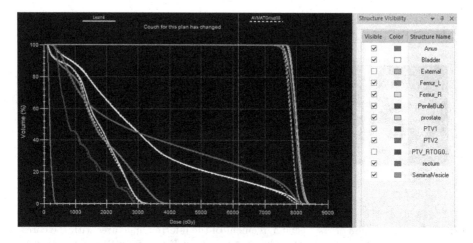

Fig. 4. Dose-volume histograms (DVHs) depict that fraction of a structure's voxels
receiving dose greater than or equal to the dose on the horizontal axis. Targets (PTV1,
PTV2) cluster at upper right and OARs occupy the lower left portion of the figure.
Two plans are compared here: a single patient's CycleGAN (solid) plan from study
4, epoch16 (Fig. 2), and the GT plan (dashed). All the OAR DVH-pairs are nearly
co-incident and the target curves agree closely.

3.1   Plan Quality Overview via Dose Volume Histograms

The cumulative dose volume histogram (DVH) depicts the fraction of a structure's voxels receiving dose greater than or equal to the dose on the horizontal
axis. Figure 4 displays the DVHs for all the targets and OARs for the minimum-
MALD Study 4, Epoch 16 model described in Figs. 2 and 3. The OAR DVHs
occupy the lower left of the figure and the target DVH curves are at the upper

right. The CycleGAN DVHs are the solid lines and the ground truth plan's DVHs are the dashed lines. In Fig. 4 the CycleGAN and ground truth OARs are nearly coincident while the target DVHs show close agreement. This pattern of OAR and target DVH agreement was observed for all the low-MALD models.

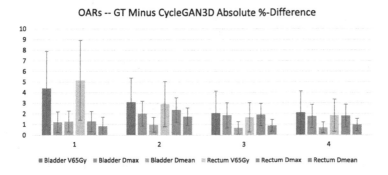

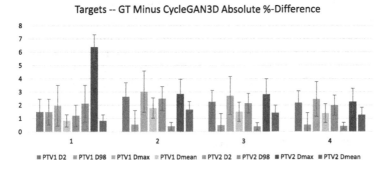

Fig. 5. Bar plots of mean absolute %-difference (GT - CycleGAN) dosimetric measures for OARs (upper) and targets (lower) for 20 test cases, for each of four learning studies. Error bars represent $\pm 1\sigma$. The OAR plots are for bladder and rectum only. V65Gy is the fraction of a structure receiving a dose greater than or equal to 65 Gy, a large dose since the prescribed target dose is 78 Gy. Dmax and Dmean are the maximum and mean doses for the structure. D98% is the maximum dose seen by target voxels receiving at least 98% of the lower doses and D2% is the minimum dose seen by 2% of the higher-dose voxels. Overall, the average absolute differences for both OARs and targets are mostly 3% or less of the GT value.

3.2 Dosimetric Properties of the Inferred Treatment Plans

Figure 5 presents another view of the Study 4, Epoch 16 inferred doses. Plotted are the mean absolute %-differences, $1/N \Sigma_{n=1}^{N}[100 * |CycleGAN_n - GT_n|/GT_n]$ for the $N = 20$ test cases, for each of a set of dose metrics, for OARs bladder and rectum (top) and targets PTV1 and PTV2 (below). (Error bars represent $\pm 1\sigma$.) Dmean and Dmax are the mean and maximum doses observed for each structure and V65Gy (%) is the volume percent of a structure receiving a dose of 65 Gray (Gy). D98% is the largest dose seen by 98% of the lower-dose target voxels and D2% is the smallest dose seen by 2% of the highest-dose target voxels. For comparison, the prescribed dose for the ground truth plans is 78 Gy.

The OAR (upper) plots show most of the dose-comparison measures at or below 3%. The trends across the series show a declining difference between the V65Gy and the Dmax and Dmean metrics, and general decline in all the measures across the four learning studies. The targets' plots (lower) present a similar picture with declining differences between Dmax and Dmean within models, and overall declines in all the measures from studies 1 to 4.

3.3   Predicted-Plan v. Predicted-Dose Results

Comparison with published work is problematic since these results and those of other investigators were obtained under different conditions. Nguyen et al. [7] presented a dose prediction study using a 2D U-Net on 88 prostate cases (training(80); test (8)). Models were learned using 10-fold cross-validation; the best performing model was used for testing. They found [7] Table 1, mean absolute Dmax and Dmean %-differences of 1.80, 1.09 (PTV), 1.94, 4.22 (bladder), and 1.26, 1.62 (rectum), respectively. The corresponding Dmax, Dmean absolute %-differences for the Study4, Epoch 16 model are: 2.45, 1.39 (PTV1), 2.26, 1.27 (PTV2), 1.76, 0.69 (bladder), and 1.80, 0.97 (rectum), respectively. At least we can say that these predicted-plan results are similar to the predicted-dose results of [7].

4   Summary and Discussion

This is the first report to our knowledge of a deep learning model for radiotherapy treatment plans realized as Linac/MLC parameters. The 3D CycleGAN network predicts the parameters for a novel-patient anatomy rendered as a 3D array of beam's-eye-view projections, and the resulting apertures and their weights can be directly input into a TPP for dose calculation and treatment delivery if desired. In general, the 3D CycleGAN network tends to slightly under-estimate the PTV doses, but the OAR sparing is nearly identical to that of the ground truth plans.

TPPs must provide a heuristic control point initialization to begin VMAT plan computation, and it would be interesting to know how the CycleGAN predicted control points compare to the heuristic control points. First, the TPP-supplied control points always have to be optimized with respect to treatment objectives requiring 30–60 min for a typical prostate planned using the criteria of this study. The CycleGAN predictions learned from a population of high quality plans are already close (on average) to the ground truth plans (see Fig. 5) and because they are re-composed as DICOM RT Plan files, they could potentially be used for treatment. When refinement is required, the CycleGAN control point refinement typically takes less time than that of the default TPP initial control points (data not shown). Studies of this and other properties of the CycleGAN predicted plans are in progress.

The 3D CycleGAN models can serve several functions. The network inferences provide a lower bound to the plan quality achievable using the plan objectives and constraints for this particular radiotherapy treatment. Thus, if a new plan in progress is not as good as the network-predicted plan, the planner knows at least that a better plan is possible, knowledge not available with current technology. Second, these inferences may to input to a TPP (e.g., Elekta Monaco) for MLC aperture shape and weight refinement for a warm start dose calculation. Either way, a clinic without deep physics expertise or with very demanding treatment throughput requirements could achieve better efficiency and plan quality.

Acknowledgements. The author would like to thank Peter Voet, Hafid Akhiat, Spencer Marshall, and Michel Moreau for help creating the prostate plans and for many helpful discussions.

References

1. Kazhdan, M., et al.: A shape relationship descriptor for radiation therapy planning. In: Yang, G.-Z., Hawkes, D., Rueckert, D., Noble, A., Taylor, C. (eds.) MICCAI 2009. LNCS, vol. 5762, pp. 100–108. Springer, Heidelberg (2009). https://doi.org/10.1007/978-3-642-04271-3_13
2. Wu, B., Ricchetti, B., Sanguinetti, G., et al.: Patient geometry-driven information retrieval for IMRT treatment plan quality control. Med. Phys. **36**(12), 5497–5505 (2009)
3. Zhu, X., Ge, Y., Li, T., Thongphiew, D., Yin, F.-F., Wu, Q.J.: A planning quality evaluation tool for prostate adaptive IMRT based on machine learning. Med. Phys. **38**(2), 719–726 (2011)
4. Appenzoller, L.M., Michalski, J.M., Thorstad, W.L., Mutic, S., Moore, K.L.: Predicting dose-volume histograms for organs-at-risk in IMRT planning. Med. Phys. **39**(12), 7446–7461 (2012)
5. Shiraishi, S., Moore, K.L.: Knowledge-based prediction of three-dimensional dose distributions for external beam radiotherapy. Med. Phys. **43**(1), 378–387 (2012)
6. McIntosh, C., Purdie, T.G.: Contextual atlas regression forests: multiple-atlas-based automated dose prediction in radiation therapy. IEEE Trans. Med. Imaging **35**(4), 1000–1012 (2016)
7. Nguyen, D., Long, T., Jia, X.: A feasibility study for predicting optimal radiation therapy dose distributions of prostate cancer patients from patient anatomy using deep learning. Sci. Rep. **9**, 1076 (2019). https://doi.org/10.1038/s41598-018-37741-x
8. Kearney, V., et al.: DoseGAN: a generative adversarial network for synthetic dose prediction using attention-gated discrimination and generation. Sci. Rep. **10**(1), 1–8 (2020)
9. Unkelbach, J., et al.: Optimization approaches to volumetric modulated arc therapy planning. Med. Phys. **42**(3), 1367–1377 (2015)
10. RTOG Homepage. http://www.rtog.org/. Accessed 10 Mar 2020
11. Breedveld, S., Storchi, P.R.M., Heijmen, B.J.M.: The equivalence of multi-criteria methods for radiotherapy plan optimization. Phys. Med. Biol. **54**, 7199–7209 (2009)

12. Breedveld, S., Storchi, P.M.R., Voet, P.W.J., Heijmen, B.J.M.: iCycle: integrated, multicriterial beam angle, and profile optimization for generation of coplanar and noncoplanar IMRT plans. Med. Phys. **39**(2), 951–963 (2012)
13. Wang, Y., Breedveld, S., Heijmen, B.J.M., Petit, S.F.: Evaluation of plan quality assurance models for prostate cancer patients based on fully automatically generated Pareto-optimal treatment plans. Phys. Med. Biol. **61**, 4268–4282 (2016)
14. Rit, S., Vila Oliva, M., Brousmiche, S., Labarbe, R., Sarrut, D., Sharp, G.C.: The Reconstruction Toolkit (RTK), an open-source cone-beam CT reconstruction toolkit based on the Insight Toolkit (ITK). J. Phys. Conf. Ser. **489**, 012079 (2013)
15. Goodfellow, I., Pouget-Abadie, J., Mirza, M., et al.: Generative adversarial nets. arXiv:1406.2661v1 (2014)
16. Goodfellow, I.: NIPS 2016 tutorial: generative adversarial networks. arXiv:1701.00160v4 (2016)
17. Zhu, J.T., Park, T., Isola, P., Efros, A.A.: Unpaired image-to-image translation using cycle-consistent adversarial networks. arXiv:1703.10593v1 (2017)
18. Isola, P., Zhu, J.-T., Zhou, T., Efros, A.A.: Image-to-image translation with conditional adversarial networks. arXiv:1611.07004v1 (2016)
19. Ulyanov, D., Vedaldi, A., Lempitsky, V.: Instance normalization: the missing ingredient for fast stylization. arXiv:1607.08022v3 (2017)
20. Ronneberger, O., Fischer, P., Brox, T.: U-net: convolutional networks for biomedical image segmentation (2015). arXiv.1505.04597v1
21. Milletari, F., Navab, N., Ahmadi, S.-A.: V-net: fully convolutional neural networks for volumetric medical image segmentation. In: Fourth International Conference on 3D Vision (3DV) (2016)
22. Ioffe, S., Szegedy, C.: Batch normalization: accelerating deep network training by reducing internal covariate shift. arXiv:1502.03167v3 (2015)
23. Kingma, D.P., Ba, J.L.: ADAM: a method for stochastic optimization. arXiv:1412.6980v8 (2015)
24. Chollet, F.: Deep Learning with Python. Manning, New York (2018)
25. Abadi, M., Agarwal, A., Barham, P., et al.: TensorFlow: Large-Scale Machine Learning on Heterogeneous Distributed Systems. Google Research (2015). https://www.tensorflow.org/
26. Brownlee, J.: Generative Adversarial Networks with Python (2019). http://machinelearningmastery.com

ML-CDS 2020

Soft Tissue Sarcoma Co-segmentation in Combined MRI and PET/CT Data

Theresa Neubauer[1], Maria Wimmer[1(✉)], Astrid Berg[1], David Major[1], Dimitrios Lenis[1], Thomas Beyer[2], Jelena Saponjski[3], and Katja Bühler[1]

[1] VRVis Zentrum für Virtual Reality und Visualisierung Forschungs-GmbH, Vienna, Austria
mwimmer@vrvis.at

[2] QIMP Team, Center for Medical Physics and Biomedical Engineering, Medical University of Vienna, Vienna, Austria

[3] Center for Nuclear Medicine, Clinical Center of Serbia, Belgrade, Serbia

Abstract. Tumor segmentation in multimodal medical images has seen a growing trend towards deep learning based methods. Typically, studies dealing with this topic fuse multimodal image data to improve the tumor segmentation contour for a single imaging modality. However, they do not take into account that tumor characteristics are emphasized differently by each modality, which affects the tumor delineation. Thus, the tumor segmentation is modality- and task-dependent. This is especially the case for soft tissue sarcomas, where, due to necrotic tumor tissue, the segmentation differs vastly. Closing this gap, we develop a modality-specific sarcoma segmentation model that utilizes multimodal image data to improve the tumor delineation on each individual modality. We propose a simultaneous co-segmentation method, which enables multimodal feature learning through modality-specific encoder and decoder branches, and the use of resource-efficient densely connected convolutional layers. We further conduct experiments to analyze how different input modalities and encoder-decoder fusion strategies affect the segmentation result. We demonstrate the effectiveness of our approach on public soft tissue sarcoma data, which comprises MRI (T1 and T2 sequence) and PET/CT scans. The results show that our multimodal co-segmentation model provides better modality-specific tumor segmentation than models using only the PET or MRI (T1 and T2) scan as input.

Keywords: Tumor co-segmentation · Multimodality · Deep learning

1 Introduction

In cancer therapy, automatic tumor segmentation supports healthcare professionals as it provides a fast quantitative description of the tumor volume and location. To analyze soft tissue sarcomas in more detail, usually, complementing imaging modalities are used to depict the tumor from an anatomical or physiological perspective, such as Magnetic Resonance Imaging (MRI), Computed

© Springer Nature Switzerland AG 2020
T. Syeda-Mahmood et al. (Eds.): ML-CDS 2020/CLIP 2020, LNCS 12445, pp. 97–105, 2020.
https://doi.org/10.1007/978-3-030-60946-7_10

Tomography (CT), or Positron Emission Tomography (PET). These modalities show different characteristics of the tumor tissue and thus provide valuable complementary information. However, depending on the imaging modality and clinical indication, the segmentation contour may look different for the same tumor.

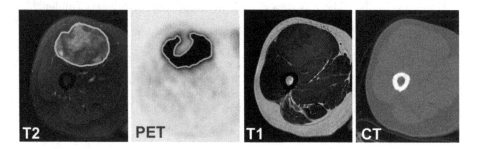

Fig. 1. Depending on the modality and the clinical intent, the segmentation for soft tissue sarcomas on the MRI T2 scan (yellow contour) and the PET scan (green contour) may look different. Figure best viewed in color. (Color figure online)

Soft tissue sarcomas are malignant tumors that originate from various tissues, including muscular tissue, connective tissue, and nervous tissue. They predominantly occur in the extremities. Due to their large size, soft tissue sarcomas tend to form necrotic tumor areas. In MRI scans, necrosis is considered part of the tumor, but it is not visible on the PET scan as the necrosis is no longer metabolically active. Figure 1 demonstrates the challenge of multimodal segmentation for soft tissue sarcomas on PET and MRI scans.

Deep learning based multimodal tumor segmentation methods have been proposed, e.g. for brain tumor segmentation on multi-sequence MRIs [3,7] or lung tumor segmentation on PET/CTs [5,12]. Current state-of-the-art networks are inspired by fully convolutional neural networks (FCNs), whereby different ways to incorporate the complementary information of multimodal image data have been presented. These multimodal segmentation studies report a better segmentation result compared to models using monomodal images. However, the main limitation of these studies is that one modality is set as the segmentation target for the final contour and thus only one modality-specific tumor volume is obtained. Contrary, in cancer therapy there are different clinical routines, which require a set of modality-specific tumor delineations from the input data.

To solve this problem for sarcomas, we aim to *simultaneously co-segment* selected modality-specific tumor volumes from the given input modalities. To the best of our knowledge, there is only the study of Zhong et al. [12], which investigates tumor co-segmentation with deep learning. They perform lung tumor segmentation on PET/CT scans, co-segmenting the modality-specific tumor in both the CT and PET scan. However, their use of two connected 3D U-Nets (one per modality), results in a very large model with more than 30M parameters.

Therefore, we introduce a resource-efficient, multimodal network for sarcoma co-segmentation, which allows the network to simultaneously segment several modality-specific tumor volumes on a subset of the input modalities. Our model benefits from (1) modality-specific encoders and decoders for multimodal feature learning, and (2) dense blocks for efficient feature re-use. We demonstrate the effectiveness of our method on public soft-tissue sarcoma data [1,10,11] and extensively evaluate the influence of MRI and PET/CT data for co-segmentation.

2   Method

For each patient i, $i = 1, \ldots, n$, let $\mathcal{I}_i$ be a set of medical images of fixed modalities corresponding to this patient, i.e. $\mathcal{I}_i := \{I_i^m\}_{i,m}$ with I_i^m an image of patient i and modality $m \in \{T1, T2, CT, PET\}$. For every $\mathcal{I}_i$, we define the set of corresponding ground truth segmentation masks $\mathcal{M}_i := \{M_i^{m'}\}_{i,m'}$ where $m' \in \{T2, PET\}$. We then seek for a co-segmentation network that is capable of estimating the given ground truth masks $\mathcal{M}_i$, given a chosen subset of input modalities. Our proposed model is inspired by the popular U-Net [9] architecture, and the work of Jégou et al. [4], who extended the DenseNet [2] for the task of semantic segmentation. Figure 2 gives an overview of our model, which comprises the following main parts:

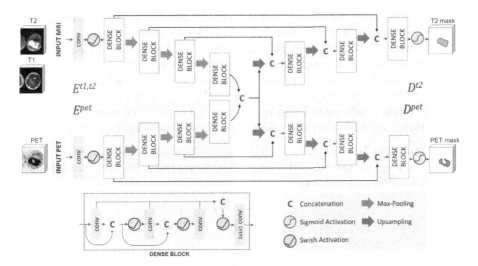

Fig. 2. We use two separated encoder branches $E^{t1,t2}$ and E^{pet} for modality-specific feature extraction and pass the concatenated latent representation to both decoders D^{t2} and D^{pet} for efficient segmentation of both tumor contours. Best viewed in color. (Color figure online)

Modality-Specific Encoder and Decoder. We use two different encoder branches $E^{t1,t2}$ and E^{pet} for MRI and PET data, respectively, to extract features for each target modality separately. In the MRI branch, we additionally use the T1 scan as a supporting modality to improve the feature extraction of the target T2 scan. The separation of the modality types in the encoder part is inspired by prior work on multimodal segmentation models [13]. Firstly, studies with multi-sequence MRIs have shown that input-level fusion leads to a significant improvement in model performance [3,7]. Secondly, for studies dealing with complementary modalities such as PET/CT, modality-specific encoder branches are widely used [5,12].

Each encoder, $E^{t1,t2}$ and E^{pet}, consists of an initial $3 \times 3 \times 3$ convolution layer with 48 filters, followed by four dense blocks. After each block, the resulting feature map is then downsampled using a max-pooling layer with pooling size $2 \times 2 \times 2$ and stride 2, which halves the resolution of the feature maps. To account for the low spatial resolution of the z-axis of the MRI scans, we propose to perform $2 \times 2 \times 1$ pooling after the second dense block instead.

We concatenate the latent representation of $E^{t1,t2}$ and E^{pet} and pass the feature maps to both decoders D^{t2} and D^{pet}. Each dense block in each decoder D^{t2} and D^{pet} receives the feature maps of the dense block at the same resolution level from the corresponding encoder $E^{t1,t2}$ and E^{pet}, respectively. In the following we refer to our proposed model as $E^{t1,t2}E^{pet}$-$D^{t2}D^{pet}$.

Dense Blocks. Each dense block consists of three repeated $3 \times 3 \times 3$ convolution layers and Swish [8] activations. This iterative concatenation of feature maps leads to feature re-use, which in turn reduces the number of parameters [2]. The number of filters of all convolution layers in a block is increased with each block level, learning 12, 28, 44, or 60 filters, respectively. In contrast to Jégou et al. [4], we removed the batch normalization layers, since we use a batch size of one. We also removed the dropout layers, because they did not lead to performance improvements. At the end of the dense block, the feature maps of all convolution layers are then concatenated and reduced by a factor of 0.5 using a $1 \times 1 \times 1$ convolution layer to further decrease the number of model parameters.

Loss Function. To account for both tumor masks in our co-segmentation model during training, we calculate the dice losses individually for each mask in $\mathcal{M}_i$ and combine them as follows:

$$\mathcal{L} = - \sum_{m' \in \{T2, PET\}} \frac{2 \mid M^{m'} \cap P^{m'} \mid + \epsilon}{\mid M^{m'} \mid + \mid P^{m'} \mid + \epsilon} \tag{1}$$

whereby $M^{m'}$ and $P^{m'}$ denote the voxel set of the ground truth volume $M^{m'}$ and the predicted volume $P^{m'}$ belonging to modality $m' \in \{T2, PET\}$. The parameter ϵ is added to avoid numerical instabilities.

Variant: Shared Decoder. We further introduce a lightweight variant of our model which uses only one shared decoder $D^{t2,pet}$. Here, each dense block receives the multiplied feature maps from the $E^{t1,t2}$ and E^{pet} encoder block at the same level. The fusion of feature maps by multiplication is intended to emphasize the overlapping position of the two masks. However, the feature maps of the first encoder blocks are fused by concatenation to allow for modality-specific differences in the segmentation masks. The last layer of the decoder has two output channels: one for the MRI mask M_i^{t2} and one for the PET mask M_i^{pet}. We compare both models in Sect. 4.

3 Experimental Setup

3.1 Dataset and Pre-processing

We evaluate our method on the soft tissue sarcoma dataset [10,11], which is publicly available at The Cancer Imaging Archive [1]. The highly heterogeneous dataset comprises 51 patients with sarcomas in the extremities, with the data coming from different sites and scanners. For each patient, four different imaging modalities have been acquired: two paired MRI (T1 and T2) scans and a PET/CT scan. The MRI and PET/CT exams were acquired on different days, resulting in changed body positions as well as anatomical variations. The dataset already includes tumor annotations, which are delineated on the T2 scans. In addition, an experienced nuclear physician delineated the tumor contours for our study on the PET scan for radiotherapy treatment. We pre-processed the dataset as follows:

- **Co-registration:** We followed Leibfarth et al. [6] for multimodal intra-patient registration and registered the PET/CT scan with the corresponding PET contour on the T2 scan.
- **Resampling:** The in-plane pixel resolution was resampled to 0.75×0.75 mm using B-Spline interpolation, while the slice distance was kept at the original distance of the T2 scan to avoid resampling artifacts due to the low spatial resolution.
- **Crop images:** We focus on patients with tumors in their legs and cropped all scans to the leg region, resulting in 39 patients. The cropped scans have varying sizes (210×210 to 600×660) and slice numbers (15 to 49).
- **Modality-dependent intensity normalization:** We applied z-score normalization to the T1 and T2 scans. The PET scans were normalized by a transformation to standard uptake values using body-weight correction.

3.2 Network Training

We randomly divide the 39 patients into five distinct sets and perform 5-fold cross-validation. We increase the efficiency of the training using 3D patches of size $256 \times 256 \times 16$, which are randomly extracted from the image while ensuring that tumor tissue is visible on every patch. To avoid overfitting and account for the small number of training samples, we perform the following data augmentation strategies: scaling, rotation, mirroring, and elastic transformations. We train our network using the loss function Eq. 1 and the Adam optimizer with a batch size of one. We start with an initial learning rate of $1e^{-4}$, which is reduced by a factor of 0.5 if the validation loss has not decreased for eight epochs. All convolutional kernels are initialized with he_normal. The models are implemented using Keras with Tensorflow backend and trained on an NVIDIA Titan RTX GPU (24 GB RAM).

3.3   Evaluation Measures

The segmentation performance is measured calculating the overlap-based dice similarity coefficient (DSC) and distance-based average symmetric surface distance (ASSD) for each predicted mask P of modality m' and its corresponding ground truth mask $M \in \mathcal{M}_i$. Formally:

$$DSC^{m'}(M, P) = \frac{2 \mid M \cap P \mid}{\mid M \mid + \mid P \mid} \tag{2}$$

$$ASSD^{m'}(M, P) = \frac{\sum_{g_k \in M} d(g_k, M) + \sum_{p_k \in P} d(p_k, P)}{\mid M \mid + \mid P \mid} \tag{3}$$

whereby $g_k \in M$ and $p_k \in P$ denote a voxel in the ground truth volume M and predicted volume P, respectively. The Euclidean distance $d(g_k, P)$ is calculated between voxel g_k and the closest voxel in P.

4   Results and Discussion

We compare the performance of our proposed network $E^{t1,t2}E^{pet}\text{-}D^{t2}D^{pet}$ with different baseline models. These experiments demonstrate the influence of varying sets of input modalities as well as modality-specific encoder/decoder designs for our model. Table 1 summarizes mean DSC (in %) and ASSD (in mm) for T2 and PET segmentation separately. Visual results are shown in Fig. 3. To compare our approach to the state-of-the-art, we implement the model by Zhong et al. [12] using two parallel U-Nets: one for the T2 scan and one for the PET scan yielding the segmentation masks for the T2 and PET scan simultaneously. We followed the proposed implementation details. However, to allow for a fair comparison, we changed the patch size to our settings. Additionally, we adapt our z-axis pooling approach to the model of Zhong et al. and name it *Zhong modified*.

Single Modality Mask Prediction: When comparing the scores for the prediction of M_i^{t2} only, we found that the lowest results are achieved when only using T2 as input. The performance increases when incorporating both T1 and T2 in the encoders, whereby the best results are obtained with a shared encoder used in model $E^{t1,t2}\text{-}D^{t2}$. These results confirm our choice for the shared MRI encoder $E^{t1,t2}$ of our proposed model. In contrast, a single PET modality is sufficient to achieve a good PET segmentation M_i^{pet}, as shown for model $E^{pet}\text{-}D^{pet}$. We further observed, that adding a separate encoder E^{ct} to the model resulted in the highest performance increase, yielding the best scores for predicting M_i^{pet} overall ($76.1\% \pm 16.0\%$ DSC, $3.7\,\text{mm} \pm 4.1\,\text{mm}$ ASSD).

Table 1. Performance metrics per model: Mean DSC and ASSD and their standard deviation calculated for the T2 and PET segmentation masks. All results were obtained by running a 5-fold cross-validation. Modalities used in the first encoder branch are denoted by •, and the ones in the second encoder branch are denoted by ∘.

Model	T1	T2	PET	CT	Mean DSC (%) T2	Mean DSC (%) PET	ASSD (mm) T2	ASSD (mm) PET
$E^{t1,t2}E^{pet}\text{-}D^{t2}D^{pet}$	•	•	∘		**77.2 ± 16.5**	74.6 ± 19.0	**3.8 ± 5.3**	4.5 ± 6.2
$E^{t1,t2}E^{pet}\text{-}D^{t2,pet}$	•	•	∘		75.3 ± 17.2	74.2 ± 19.9	4.5 ± 5.3	4.3 ± 5.4
$E^{t1,t2}E^{pet}\text{-}D^{t2}$	•	•	∘		76.5 ± 16.6		3.9 ± 4.9	.
$E^{t1,t2}E^{pet}\text{-}D^{pet}$	•	•	∘			74.9 ± 16.1	.	4.3 ± 5.1
$E^{t2}\text{-}D^{t2}$		•			65.6 ± 24.0		10.2 ± 10.9	.
$E^{t1,t2}\text{-}D^{t2}$	•	•			71.0 ± 23.8		6.5 ± 8.2	.
$E^{t1}E^{t2}\text{-}D^{t2}$	•	∘			68.3 ± 20.0		7.9 ± 8.9	.
$E^{pet}\text{-}D^{pet}$			•		.	74.3 ± 18.8	.	4.9 ± 6.4
$E^{pet,ct}\text{-}D^{pet}$			•	•	.	74.4 ± 21.6	.	5.5 ± 14.1
$E^{pet}E^{ct}\text{-}D^{pet}$			•	∘	.	**76.1 ± 16.0**	.	3.7 ± 4.1
$E^{t2}E^{pet}\text{-}D^{t2}D^{pet}$		•	∘		72.1 ± 19.8	73.5 ± 20.0	4.9 ± 5.3	4.8 ± 6.2
$E^{t2}E^{pet}\text{-}D^{t2,pet}$		•	∘		71.6 ± 19.4	73.2 ± 19.5	5.6 ± 6.0	4.2 ± 5.1
Zhong [12]		•	∘		72.4 ± 20.0	74.1 ± 19.4	6.7 ± 8.8	4.0 ± 4.9
Zhong modified		•	∘		75.6 ± 16.2	75.2 ± 17.2	4.8 ± 6.5	**3.6 ± 4.3**

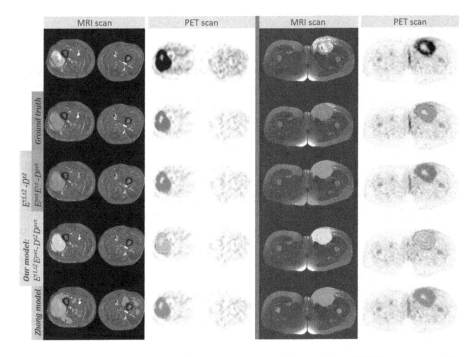

Fig. 3. Visual segmentation results of compared models on T2 and PET scan pairs. The presented samples confirm the trend from Table 1: The variation of the predicted T2 masks is higher between different models, while the impact for the PET segmentations is less apparent. Best viewed in color. (Color figure online)

Encoder/Decoder Design: The results in Table 1 suggest that the segmentation performance benefits from modality-specific encoders that separate anatomical and functional modalities. Comparing the models with shared $D^{t2,pet}$ and separate decoders $D^{t2}D^{pet}$, we report lower DSC scores when using the proposed shared decoder variant of our model. The performance impact is higher for T2, which is also reflected by DSC and ASSD scores.

Co-segmentation: Looking at the sarcoma co-segmentation models, we observe that the tumor delineation on MRI T2 scans benefits from the feature co-learning with PET. This is reflected by the best overall scores (77.2% ± 16.5% DSC, 3.8 mm ± 5.3 mm ASSD) obtained with our proposed model $E^{t1,t2}E^{pet}$-$D^{t2}D^{pet}$. The same is observed with model $E^{t1,t2}E^{pet}$-D^{t2} for predicting only the T2 mask, which gives comparable results to our model. Model $E^{t1,t2}E^{pet}$-$D^{t2}D^{pet}$ outperforms the method by Zhong et al. [12] and achieves similar DSC and ASSD values to the model *Zhong modified*. However, our $D^{t2,pet}$ models (max. 3.9M) and $D^{t2}D^{pet}$ models (max. 5.5M) require - depending on the encoder - only 10–16% of the parameters of Zhong et al. (33.7M) and are therefore much more resource-efficient. When comparing the model of Zhong et al. with *Zhong modified*, it is revealing that the adaption of the pooling strategy to the anisotropic data resolution yields notable performance gains.

5   Conclusion

In this paper, we proposed a simultaneous co-segmentation model for soft tissue sarcomas, which utilizes densely connected convolutional layers for efficient multimodal feature learning. We performed an extensive evaluation, comparing various ways to incorporate multimodal data (MRI T1 and T2, CT and PET) into our model. We showed that our proposed network outperforms the state-of-the-art method for tumor co-segmentation, yielding better or comparable results for MRI T2 and PET, respectively. Moreover, our proposed co-segmentation architecture and single-modal variants reduce the number of parameters by up to 90% compared to the concurring method. These experiments show (1) improved accuracy when using multimodal data and (2) demonstrate that the choice of input modalities and encoder-decoder architecture is crucial for the segmentation result.

Acknowledgements. VRVis is funded by BMK, BMDW, Styria, SFG and Vienna Business Agency in the scope of COMET - Competence Centers for Excellent Technologies (854174) which is managed by FFG.

References

1. Clark, K., et al.: The cancer imaging archive (TCIA): maintaining and operating a public information repository. J. Digit. Imaging **26**(6), 1045–1057 (2013). https://doi.org/10.1007/s10278-013-9622-7

2. Huang, G., Liu, Z., Van Der Maaten, L., Weinberger, K.Q.: Densely connected convolutional networks. In: Proceedings of the IEEE CVPR, pp. 4700–4708. IEEE (2017)

3. Isensee, F., Kickingereder, P., Wick, W., Bendszus, M., Maier-Hein, K.H.: No new-net. In: Crimi, A., Bakas, S., Kuijf, H., Keyvan, F., Reyes, M., van Walsum, T. (eds.) BrainLes 2018. LNCS, vol. 11384, pp. 234–244. Springer, Cham (2019). https://doi.org/10.1007/978-3-030-11726-9_21

4. Jégou, S., Drozdzal, M., Vazquez, D., Romero, A., Bengio, Y.: The one hundred layers tiramisu: fully convolutional DenseNets for semantic segmentation. In: Proceedings of the IEEE CVPR Workshops, pp. 11–19. IEEE (2017)

5. Kumar, A., Fulham, M., Feng, D., Kim, J.: Co-learning feature fusion maps from PET-CT images of lung cancer. IEEE Trans. Med. Imaging 39(1), 204–217 (2020)

6. Leibfarth, S., et al.: A strategy for multimodal deformable image registration to integrate PET/MR into radiotherapy treatment planning. Acta Oncologica 52(7), 1353–1359 (2013)

7. Myronenko, A.: 3D MRI brain tumor segmentation using autoencoder regularization. In: Crimi, A., Bakas, S., Kuijf, H., Keyvan, F., Reyes, M., van Walsum, T. (eds.) BrainLes 2018. LNCS, vol. 11384, pp. 311–320. Springer, Cham (2019). https://doi.org/10.1007/978-3-030-11726-9_28

8. Ramachandran, P., Zoph, B., Le, Q.V.: Searching for activation functions. arXiv preprint arXiv:1710.05941 (2017)

9. Ronneberger, O., Fischer, P., Brox, T.: U-Net: convolutional networks for biomedical image segmentation. In: Navab, N., Hornegger, J., Wells, W.M., Frangi, A.F. (eds.) MICCAI 2015. LNCS, vol. 9351, pp. 234–241. Springer, Cham (2015). https://doi.org/10.1007/978-3-319-24574-4_28

10. Vallières, M., Freeman, C.R., Skamene, S.R., El Naqa, I.: A radiomics model from joint FDG-PET and MRI texture features for the prediction of lung metastases in soft-tissue sarcomas of the extremities. Cancer Imaging Arch. (2015). https://doi.org/10.7937/K9/TCIA.2015.7GO2GSKS

11. Vallières, M., Freeman, C.R., Skamene, S.R., El Naqa, I.: A radiomics model from joint FDG-PET and MRI texture features for the prediction of lung metastases in soft-tissue sarcomas of the extremities. Phys. Med. Biol. 60(14), 5471–5496 (2015)

12. Zhong, Z., et al.: Simultaneous cosegmentation of tumors in PET-CT images using deep fully convolutional networks. Med. Phys. 46(2), 619–633 (2019)

13. Zhou, T., Ruan, S., Canu, S.: A review: deep learning for medical image segmentation using multi-modality fusion. Array 3–4(10004), 1–11 (2019)

Towards Automated Diagnosis with Attentive Multi-modal Learning Using Electronic Health Records and Chest X-Rays

Tom van Sonsbeek[(✉)] and Marcel Worring

University of Amsterdam, Amsterdam, The Netherlands
{t.j.vansonsbeek,m.worring}@uva.nl

Abstract. Jointly learning from Electronic Health Records (EHR) and medical images is a promising area of research in deep learning for medical imaging. Using the context available in EHR together with medical images can lead to more efficient data usage. Recent work has shown that jointly learning from EHR and medical images can indeed improve performance on several tasks. Current methods are however still not independent of clinician input. To obtain an automated method only prior patient information should be used together with a medical image, without the reliance on further clinician input. In this paper we propose an automated multi-modal method which creates a joint feature representation based on prior patient information from EHR and associated X-ray scan. This feature representation, which joins the two different modalities through attention leverages the contextual relationship between the modalities. This method is used to perform two tasks: diagnosis classification and free-text diagnosis generation. We show the benefit of the multi-modal approach over single-modality approaches on both tasks.

Keywords: Electronic Health Records · Multi-modal learning · Deep learning

1 Introduction

Deep learning models are by a large distance the top-performing algorithms on most medical imaging tasks, and even outperform specialist clinicians on some specific tasks [12]. One of the key factors driving the performance of deep learning methods is the quantity of available data. While generally more data leads to better performance [16], it is not always possible to obtain large (annotated) datasets for medical image analysis [3]. Therefore, it is of utmost importance to optimally use available data.

Most current medical imaging methods only make use of the scans from imaging studies. Patient history, notes of clinicians, etc., recorded in the Electronic Health Records (EHR) associated to the scan are not often used. It is shown

© Springer Nature Switzerland AG 2020
T. Syeda-Mahmood et al. (Eds.): ML-CDS 2020/CLIP 2020, LNCS 12445, pp. 106–114, 2020.
https://doi.org/10.1007/978-3-030-60946-7_11

that (longitudinal) EHR data can be used as a diagnosis predictor [2,5,18,21]. Combined usage of EHR and medical images might be able to improve the performance of various automated medical imaging tasks to support clinicians.

The reason why using EHR and images together in the medical imaging field is not yet common practise can be attributed to multiple factors. Firstly, representing EHR in models is complicated by the large variety in content and structure, language, noise, random errors, and sparseness. Furthermore there are privacy concerns in using medical images associated with (longitudinal) EHR data, as their combined use limits the extent of possible anonymization [20]. Finally, combination of visual and textual modalities adds complexity as these methods span both image and language processing fields.

The structural difference between EHR and medical images is a major technical challenge when creating a joint learning model using these two modalities. To perform a task in a multi-modal model, there needs to be interaction between the two modalities, i.e. data fusion [22]. There are two distinct types of data fusion. Merging of modalities immediately after feature extraction is called early fusion. In early fusion modalities are integrated within the model, where inter-modality interactions are prioritized. Merging modalities in the final part of a model is called late fusion. In this type of fusion uni-modal model branches are combined thus intra-modality interactions are prioritized. A variant in between, or a combination of these two fusion types is called intermediate fusion [22].

The current largest public dataset containing linked EHR and medical images is the MIMIC-CXR dataset [8]. This dataset shows a clear structuredness of EHR in 3 parts. The first part (indication) contains information about the patient, known conditions and other information known prior to scanning of the medical image. The second part, the findings, contain observations based on the medical image, and the last part, called impression contains the diagnosis of the clinician.

A clinically useful autonomous method to support clinician decision making, which jointly learns from EHR and medical images, should only require information from the EHR available before the medical image is captured. In this way a diagnosis can be predicted without the need for additional clinician input: the multi-modal method would predict the conclusions of the clinician. In context of the structure of the EHR outlined above this would mean just the indication section of the EHR and accompanying medical images should be input for a fully autonomous multi-modal model predicting diagnosis.

1.1   Related Work

Recent work on applying deep learning techniques to jointly learn from EHRs and medical images by Tian et al. [17] showed how using a part of a structured EHR (indication and findings sections) in combination with the corresponding chest X-ray image can generate a diagnosis. In this method co-attention was used to jointly learn from CNN image features and BiLSTM features from text reports. Wang et al. [19] showed how jointly learning from full EHRs and chest X-ray images is beneficial for classification performance, and how synthetically generated reports based on radiology images improves classification performance compared to image-only models.

Nunes et al. [13] created a multi-label classification model, with two branches, which learns based on a full EHR and chest X-rays separately and merges those in the final classification layer. EHR are encoded using BioWordVec embeddings, followed by multiplicative LSTMs and multihead self-attention.

These studies show the benefit of a combined learning approach using EHR and medical images and the variety in possible downstream tasks. The input requirement of these methods, however extends beyond just the indication section the EHR. The findings section or even the full EHR as model input is required as model input. These sections of the EHR are created by the clinician after analysis of the medical image. These methods can therefore not be used as a fully automated method.

2   Proposed Approach

We propose a multi-modal method which requires the indication section of an EHR and its associated X-ray scan. As indicated, methods for automated diagnosis should only use prior patient information available before the medical image is taken. No additional input after capturing of the medical image is needed, which means this method can perform downstream tasks fully automated to support clinicians. In this multi-modal method features are extracted from both modalities and fusion happens through an intermediate fusion mechanism (Fig. 1). To capture the contextual relationship between textual and visual features, they are passed through attention layers in which visual features attend textual features and vice versa. Finally, the attended features are concatenated in the joint feature space.

Multi-modal features in the joint feature space are used for a diagnosis classification task and a free text diagnosis generation task which are described in more detail below. Visual and textual features are obtained through commonly-used methods to further emphasize on the effect of multi-modal learning. The code for this method is publicly available.[1]

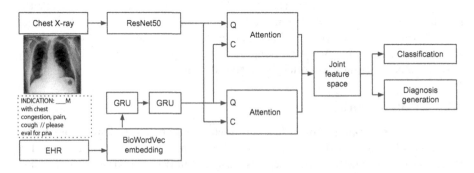

Fig. 1. Proposed multi-modal architecture.

---

[1] https://github.com/tjvsonsbeek/Multi-modal-automated-diagnosis-with-chestXray-and-EHR.

Text Features. Text in an EHR is fundamentally different in terminology compared to most text in other NLP applications. Therefore the EHR is encoded using BioWordVec embeddings [23]. These word embeddings are pretrained on publicly available biomedical corpora, such as PubMed abstracts and clinical notes from the MIMIC-III dataset [9], a dataset consisting of EHR. A maximum number of words d_w is set. Text features are obtained by passing these word embeddings $E_i = \sum_j^{d_w} E_j \in \mathbb{R}^{d_w \times 200}$ encoded through two sequential bidirectional Gated Recurrent Units (GRUs) [4] with hidden size d_g, shown in Eq. 1, in which σ is the sigmoid function, h_j the current hidden state, h_{j-1} the previous hidden state, $\tilde{h}_j$ the candidate hidden state and b, U and W learnable parameters. The final output hidden state of the bidirectional GRU can be obtained by concatenating the final forward hidden state $\overrightarrow{h}_j$ and final backward hidden state $\overleftarrow{h}_j$: $R_i = [\overrightarrow{h}_j, \overleftarrow{h}_j] \in \mathbb{R}^{d_w \times d_g}$.

$$
\begin{aligned}
r_j &= \sigma(W_r \cdot E_{ij} + U_r \cdot h_{j-1} + b_r) \\
z_j &= \sigma(W_z \cdot E_{ij} + U_z \cdot h_{j-1} + b_z) \\
\tilde{h}_j &= tanh(W_h \cdot E_{ij} + r_j \times (U_h \cdot h_{j-1}) + b_h) \\
h_j &= (1 - z_j) \times h_{j-1} + z_j \times \tilde{h}_j
\end{aligned}
\tag{1}
$$

Visual Features. Image features are extracted using an ImageNet pre-trained ResNet-50 architecture [6]. This commonly used architecture has been proven to work in medical context [15]. The medical images are scaled down to size $(224, 224)$ before being passed through the ResNet-50 model. Following Raghu et al. [15], no fine-tuning is applied to the pre-trained architecture. The layer before the final pooling layer is used after being passed through a fully connected layer:

$$
I_i = W_I I_i^r + b_I \in \mathbb{R}^{49 \times 2048}
\tag{2}
$$

Multi-modal Fusion. The visual and textual features are passed through two multi-modal attention layers to incorporate their contextual relationship, as per Eq. 3, where query Q attends to context C. These attention layers perform attention on textual features with the visual features and vice versa. To obtain a final joint feature representation image attended text features $A_{I_i;R_i}$ and text attended image features $A_{R_i;I_i}$ are concatenated in a multi-modal fusion layer $F_i = [g(I_i, R_i), g(R_i, I_i)] \in \mathbb{R}^{d_a}$. This joint feature representation F_i can be used in downstream tasks, such as classification, segmentation, text generation, etc.

$$
\begin{aligned}
\alpha_{QC} &= \sum_s \frac{exp(Q \cdot \bar{C}_s)}{\sum_{s'=1}^{S} exp(Q \cdot \bar{C}_{s'})} C_s \\
A_{Q;C} &= g(Q, C) = tanh(W_a \cdot (\alpha_{QC} + C) + b_a)
\end{aligned}
\tag{3}
$$

3   Data and Experimental Setup

The MIMIC-CXR dataset is the largest publicly available dataset containing full-text structured EHR and accompanying annotated chest X-rays [8]. The dataset contains 377,110 chest x-rays associated with 227,827 anonymised EHR. Each EHR is associated to (multiple) frontal and/or saggital X-ray views, each labelled according to the classes in Table 1.

Table 1. Classes in MIMIC-CXR and class distribution after pruning.

Label	#
No finding	85,243
Enlarged cardiomediastinum	5,337
Cardiomegaly	33,290
Lung opacity	40,270
Lung lesion	5,256
Edema	23,558
Consolidation	9,146
Pneumonia	13,669
Atelectasis	35,467
Pneumothorax	8,272
Pleural effusion	43,428
Pleural other	1,547
Fracture	35,69
Support devices	50,809
Total dataset size	**201,733**

In the majority of cases there is only a frontal X-ray view available. To ensure dataset uniformity only frontal X-ray views are selected and an EHR is required to contain at least 1) information prior to the X-ray (indication) and 2) presence of one-sentence conclusion of the report (impression). This selection yields a total of 201,733 X-ray - EHR pairs, with a class distribution listed in Table 1. 161,387 pairs are used for training and 40,346 pairs are used for testing.

3.1   Tasks

Using the multi-modal joint features F_i from the model described above two separate tasks are performed. 1) **Diagnosis classification:** By passing the joint modality features through 3 fully connected layers, followed by a final sigmoid activation each image-EHR pair is classified into one of the 14 classes mentioned above. 2) **Free-text diagnosis generation:** Based on each image-report pair the impression section of the report is predicted using a LSTM [7] with hidden size d_l, for a maximum number of words d_o. The joint modality feature representation is passed through a LSTM with as output a sequence of BioWordVec encoded words. Performance of this task is measured in terms of BLEU [14], ROUGE-L [11] and METEOR [1].

3.2 Experimental Setup

The maximum number of words in the indication input of the model is set to $d_w = 48$. The hidden size of the GRU layers $d_g = 1024$ and the dimension of the attention blocks $d_a = 2048$.

The fully connected layers leading to the classification result have sizes {4096, 512, 14}. The generated diagnosis has a maximum number of words $d_o = 24$ and the hidden size of the LSTM for diagnosis generation $d_l = 1024$. Weighted Mean Squared Error (MSE) loss and weighted Categorical CrossEntropy loss (CE) are used as loss functions for the classification and text generation task respectively. For both tasks the model is optimized using the Adam optimizer [10] with an adaptive learning rate ranging between 10^{-4} and 10^{-6}. Training of the model is ended when there is no improvement in validation loss for 3 consecutive epochs.

4 Results

To compare the performance of our multi-modal architecture the classification and diagnosis generation tasks are also evaluated based on the single-modal visual and textual features before the attention layers of the model.

Classification results (Table 2) show the benefit of the multi-modal architecture, especially macro scores of the metrics are heavily improved in comparison to the single modality methods. The large difference between the micro and macro scores of especially the precision and recall metrics can be attributed to the large class imbalance. There are multiple classes that occur in less than 5% of the dataset.

Table 2. Classification performance of proposed multi-modal and single modal models.

	Accuracy	Precision		Recall		AUCROC	
		Micro	Macro	Micro	Macro	Micro	Macro
Only images	0.892	0.305	0.128	0.666	0.275	0.863	0.784
Only EHR	0.896	0.349	0.174	0.685	0.340	0.876	0.830
Multi-modal	**0.902**	**0.428**	**0.223**	**0.690**	**0.432**	**0.8926**	**0.844**

To validate our model performance on the classification task the model was also evaluated with extended data input as used by Wang et al. [19] and Nunes et al. [13], which showed comparable performance.

Table 3. Free text diagnosis generation performance of proposed multi-modal and single modal models.

	BLEU-1	BLEU-2	BLEU-3	BLEU-4	ROUGE	METEOR
Only images	0.285	0.189	0.124	0.0618	0.271	0.283
Only EHR	0.301	0.193	0.135	0.0732	0.298	0.290
Multi-modal	**0.357**	**0.235**	**0.164**	**0.0925**	**0.336**	**0.318**

Results on the text generation task (Table 3) resonate with the results seen in the classification task, as again the multi-modal architecture outperforms the single modality variants. Figure 2b. In general, the frequent diagnoses are correctly predicted, as well as the topic of the diagnosis. The current metrics do not take synonyms into account, which lowers the performance values reported by the metrics. Fail-cases occur when the diagnosis did not occur frequently in the training set. This expresses itself in incomplete or incorrect diagnoses. These results show that free-text diagnosis generation given our set of input data is still a challenging task.

Visualization of the attention weights for both modalities is shown in Fig. 2a. Keywords that are indicative for or represent a disease have higher textual attention weights than for instance linking words. Visual attention shows to focus around the lung area, which is in line with the expectations because the majority of diagnoses is lung-related.

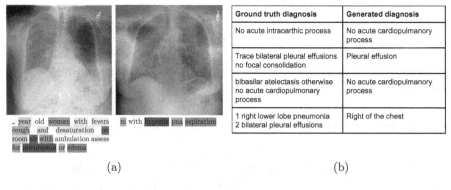

(a)   (b)

Fig. 2. Examples of model output. a) Intermediate model output: attention weights of X-ray scan attended by prior patient information and vice versa b) Comparison of generated diagnosis and ground truth diagnosis. The first three examples show approximately correct predictions, the last one shows a clear failure of the task.

5   Conclusions and Future Work

We proposed a multi-modal learning approach which leverages prior patient information from EHR and X-ray scans and their contextual relationship on two separate tasks: diagnosis classification and free text diagnosis generation. We showed the benefit of multi-modal learning compared to single-modal approaches.

This automated approach is a replication of the process in which a clinician derives conclusions from an image, given only prior information on the patient before the scan. Contrary to related methodologies this method does not require clinician input after the scan, only prior patient information is used.

In future work there are opportunities for improving the performance on both automated diagnosis tasks addressed in this paper. Incorporating more elaborate prior longitudinal information on patients has proven to be beneficial in single-modal EHR applications [5,21]. Extending this to the multi-modal domain could improve diagnosis prediction performance further as it will allow the model to learn longer term patterns from EHRs.

References

1. Banerjee, S., Lavie, A.: METEOR: an automatic metric for MT evaluation with improved correlation with human judgments. In: Proceedings of the ACL Workshop on Intrinsic and Extrinsic Evaluation Measures for Machine Translation and/or Summarization, pp. 65–72. ACL, June 2005
2. Cai, Q., Wang, H., Li, Z., Liu, X.: A survey on multimodal data-driven smart healthcare systems: approaches and applications. IEEE Access **7**, 133583–133599 (2019)
3. Cheplygina, V., de Bruijne, M., Pluim, J.P.: Not-so-supervised: a survey of semi-supervised, multi-instance, and transfer learning in medical image analysis. Med. Image Anal. **54**, 280–296 (2019)
4. Chung, J., Gulcehre, C., Cho, K., Bengio, Y.: Empirical evaluation of gated recurrent neural networks on sequence modeling. arXiv preprint arXiv:1412.3555 (2014)
5. Harerimana, G., Kim, J.W., Yoo, H., Jang, B.: Deep learning for electronic health records analytics. IEEE Access **7**, 101245–101259 (2019)
6. He, K., Zhang, X., Ren, S., Sun, J.: Deep residual learning for image recognition. In: The IEEE Conference on Computer Vision and Pattern Recognition (CVPR), June 2016
7. Hochreiter, S., Schmidhuber, J.: Long short-term memory. Neural Comput. **9**(8), 1735–1780 (1997)
8. Johnson, A.E., et al.: MIMIC-CXR: a large publicly available database of labeled chest radiographs. arXiv preprint arXiv:1901.07042 (2019)
9. Johnson, A.E., et al.: MIMIC-III, a freely accessible critical care database. Sci. Data **3**, 160035 (2016)
10. Kingma, D.P., Ba, J.: Adam: a method for stochastic optimization. arXiv preprint arXiv:1412.6980 (2014)
11. Lin, C.Y.: ROUGE: a package for automatic evaluation of summaries. In: Text Summarization Branches Out, pp. 74–81. ACL, July 2004

12. Liu, X., et al.: A comparison of deep learning performance against health-care professionals in detecting diseases from medical imaging: a systematic review and meta-analysis. Lancet Digit. Health **1**(6), e271–e297 (2019)
13. Nunes, N., Martins, B., André da Silva, N., Leite, F., Silva, M.J.: A multi-modal deep learning method for classifying chest radiology exams. In: Moura Oliveira, P., Novais, P., Reis, L.P. (eds.) EPIA 2019. LNCS (LNAI), vol. 11804, pp. 323–335. Springer, Cham (2019). https://doi.org/10.1007/978-3-030-30241-2_28
14. Papineni, K., Roukos, S., Ward, T., Zhu, W.J.: BLEU: a method for automatic evaluation of machine translation. In: Proceedings of the 40th Annual Meeting of the Association for Computational Linguistics, pp. 311–318. ACL, July 2002
15. Raghu, M., Zhang, C., Brain, G., Kleinberg, J., Bengio, S.: Transfusion: understanding transfer learning for medical imaging. Technical report (2019)
16. Sun, C., Shrivastava, A., Singh, S., Gupta, A.: Revisiting unreasonable effectiveness of data in deep learning era. In: The IEEE International Conference on Computer Vision (ICCV), October 2017
17. Tian, J., Zhong, C., Shi, Z., Xu, F.: Towards automatic diagnosis from multimodal medical data. In: Suzuki, K., et al. (eds.) ML-CDS/IMIMIC -2019. LNCS, vol. 11797, pp. 67–74. Springer, Cham (2019). https://doi.org/10.1007/978-3-030-33850-3_8
18. Tobore, I., et al.: Deep learning intervention for health care challenges: some biomedical domain considerations (2019)
19. Wang, X., Peng, Y., Lu, L., Lu, Z., Summers, R.M.: TieNet: text-image embedding network for common thorax disease classification and reporting in chest x-rays. In: Proceedings of the IEEE Conference on Computer Vision and Pattern Recognition, pp. 9049–9058 (2018)
20. Weiskopf, N.G., Hripcsak, G., Swaminathan, S., Weng, C.: Defining and measuring completeness of electronic health records for secondary use. J. Biomed. Inform. **46**(5), 830–836 (2013)
21. Xiao, C., Choi, E., Sun, J.: Opportunities and challenges in developing deep learning models using electronic health records data: a systematic review. J. Am. Med. Inform. Assoc. **25**(10), 1419–1428 (2018)
22. Zhang, C., Yang, Z., He, X., Deng, L.: Multimodal Intelligence: Representation Learning, Information Fusion, and Applications. arXiv preprint arXiv:1911.03977, November 2019
23. Zhang, Y., Chen, Q., Yang, Z., Lin, H., Lu, Z.: BioWordVec, improving biomedical word embeddings with subword information and mesh. Sci. Data **6**(1), 1–9 (2019)

LUCAS: LUng CAncer Screening with Multimodal Biomarkers

Laura Daza[1]([✉]), Angela Castillo[1], María Escobar[1], Sergio Valencia[2], Bibiana Pinzón[2], and Pablo Arbeláez[1]

[1] Center for Research and Formation in Artificial Intelligence, Universidad de los Andes, Bogotá, Colombia
{la.daza10,a.castillo13,mc.escobar11,pa.arbelaez}@uniandes.edu.co
[2] Fundación Santa Fe de Bogotá, Bogotá, Colombia

Abstract. We present the LUng CAncer Screening (LUCAS) Dataset for evaluating lung cancer diagnosis with both imaging and clinical biomarkers in a realistic screening setting. We extract key information from anonymized clinical records and radiology reports, and we use it as a natural complement to low-dose chest CT scans of patients. We formulate the task as a detection problem and we develop a deep learning baseline to serve as a future reference of algorithmic performance. Our results provide solid empirical evidence for the difficulty of the task in the LUCAS Dataset and for the interest of including multimodal biomarkers in the analysis. All the resources of the LUCAS Dataset are publicly available.

Keywords: Early lung cancer diagnosis · Multimodal biomarkers · Multimodal dataset · Lung nodules

1 Introduction

Lung cancer is the second most common type of cancer in the world [11]. Its high incidence and multiple risk factors make it a critical worldwide health problem and the focus of a considerable amount of research. In the last decade, constant progress in the development of pharmaceutical molecules and treatments for lung cancer [8] have ensured that, if detected early, the prognosis is relatively benign and the survival rate high. For instance, recent studies incorporate genetic analysis for cancer progress examination [1]. The aim was to process patient's DNA to identify the recurrence of lung cancer and find the correlation of lung nodules volume with circulating tumor DNA.

However, despite these advances, lung cancer is still the leading cause of death by cancer in the world, surpassing the combined casualties of the next three types of cancer. Every year, around 2.09 million new cases are detected (11.6% of total cases [11]) and nearly 1.76 million people die because of this disease, representing 18.4% of the total deaths by cancer [11]. Furthermore, the survival rate after five years of diagnosis for lung cancer is only 12% worldwide

© Springer Nature Switzerland AG 2020
T. Syeda-Mahmood et al. (Eds.): ML-CDS 2020/CLIP 2020, LNCS 12445, pp. 115–124, 2020.
https://doi.org/10.1007/978-3-030-60946-7_12

[24], which means that, in practice, being diagnosed today with lung cancer amounts to a death penalty for patients. The reason for this staggering mortality rate is the near-absolute absence of apparent symptoms in patients of early lung cancer [10]. Typically, symptoms become evident only when the disease is highly advanced and other organs are already compromised. Consequently, the vast majority of lung cancers worldwide are diagnosed in stages III and IV, when the efficacy of existing treatments and hence the chances of survival are seriously compromised.

To improve the prognosis in patients with lung cancer, it is necessary to make the proper early diagnosis, which in turn requires an accurate understanding of medical images because of the visual evidence of this pathology. Nowadays, physicians use Computed Tomography (CT) scans to visualize the lungs of a patient and look for any life-threatening abnormality. However, because a pulmonary nodule is a rounded or irregular opacity that measures ≤ 30 mm in diameter [14], finding them in a 3D medical image is a challenging task. Nodule detection by visual inspection is highly prone to error, even for specialists, resulting in the loss of between 43% and 52% of the nodules when evaluating the diagnostic images [18]. Once the nodules have been detected, the next step is the malignancy prediction of the findings. Specialists perform this task by visually inspecting the lesions and classifying them according to standardized descriptions of morphological characteristics, hence relying heavily on their own experience. Discrepancies among specialists are particularly concerning for this disease [6], as a single undetected or incorrectly classified malign nodule can compromise the life of the patient. This situation has spurred the appearance of machine learning techniques to assist specialists in early lung cancer detection. Accurate automated methods to perform this task would reduce the variations assessments made by different experts, providing a more robust measure of lung nodule presence, which is critical for early cancer diagnosis and treatment planning [12,22].

Progress in automated lung nodule detection and lung cancer diagnosis in the last decade is the result of a collective effort by a growing research community, which has undertaken the task of collecting and releasing large annotated datasets to train and evaluate quantitatively automated systems. A first pioneering effort was the LIDC/IDRI Database [6], which provided combined annotations by four experts of detected nodules for more than 1000 patients. Subsequently, the same data was used for the LUng Nodule Analysis (LUNA) Challenge in 2016 [16]. Concerning nodule classification, the first public challenge was the LUNGx SPIE-AAPM-NCI Lung Nodule Classification Challenge [5] in 2015, which provided malign/benign annotations for 60 nodules. In 2017, the Lung Nodule Malignancy Challenge [17] provided 1384 cases for nodule classification. In 2018, the ISBI Lung Nodule Malignancy Prediction Challenge [9] provided sequential low-dose CT (LDCT) scans at two screening intervals from the National Lung Screening Trial (NLST), with matched identified nodules from the same subject, for 100 patients. Finally, the 2020 Grand Challenge on Automatic Lung Cancer Management (LNDB) [19] focused on automatic classification of chest CT scans according to the 2017 Fleischner society pulmonary nodule guidelines for patient follow-up recommendation on 294 cases.

The availability of large-scale annotated datasets has spurred the development of deep learning techniques for nodule detection and lung cancer prediction. The current state-of-the-art is held by Ardila *et al.* [4], who obtained an overall 94.4% AUC on a National Lung Cancer Screening Trial test set with over 6000 patients. On the other hand, the best result obtained in the ISBI 2018 Lung Nodule Malignancy Prediction Challenge [9] was by [20], which obtains an AUC of 91.3% on the test set. Besides, [2] present a framework to detect lung nodules in four stages. Each stage allowed the refinement of the region to find a positive candidate and classify it as a nodule. Some other methods belong to the semi-supervised approach providing an interactive solution to the physician [7].

Although deep learning methods have pushed forward automated early lung cancer diagnosis in recent years, this task is still far from being solved and large-scale low-dose screening of the population is still years away from deployment. One of the main limitations for realistic lung cancer prediction of existing experimental frameworks is the formulation of the task itself, as all existing datasets and challenges seek to diagnose the disease *using exclusively visual data*. State-of-the-art approaches for lung cancer classification take as input a chest CT and produce a probability of cancer. This setup is radically different from clinical practice, which aggregates naturally multimodal information. Even though specialists evaluate by visual inspection standardized morphological characteristics of nodules for their classification, they also take into consideration all their knowledge of the context and the patient's history. As an example, a radiologist will not study in the same way an image from a healthy child and one from a person with a 20 pack-year smoking history.

In this paper, we present the **LUng CAncer Screening (LUCAS) Dataset**, the first multimodal experimental framework for early lung cancer diagnosis. We collected a large dataset with low-dose chest CT scans of 830 patients in a real-world screening scenario in which only a small fraction of the cases is diagnosed with lung cancer, and the rest, though sane, belong to a population that is exposed to risk factors. We complement this visual information with anonymized clinical data and additional information from the radiologists' reports. The goal of the LUCAS dataset is to serve as a testbed to assess the relative importance and complementarity of the different modalities of data for lung cancer diagnosis.

In order to assess the difficulty of the task in the LUCAS Dataset and to set a baseline for future reference, we develop a deep learning technique that combines visual and clinical data for lung cancer prediction. Given the highly unbalanced nature of the detection task we address, we model it as a detection problem and we evaluate our results with the point that maximizes the Precision-Recall curve (F-score) [13,15], a standard metric in computer vision [3] that is more stringent than the AUC-ROC used by existing datasets.

Our results show that both modalities are complementary for an accurate diagnosis. Furthermore, benchmarking results in a realistic setup with the F-score reveals the true complexity of the task, as the performance of our multimodal system on the test set is only 25%. This sobering result implies that automated lung cancer screening is still a challenging open problem. To promote the appearance of a new wave of multimodal automated methods for lung cancer diagnosis, we make publicly available all the resources of this project[1].

2 Lung Cancer Screening (LUCAS) Dataset

To create the LUCAS dataset, we partnered with a large healthcare institution that provides treatment to patients with a wide variety of diseases and with different risks of lung cancer. For the data acquisition process, we first collect all the chest CT scans that had been done in the healthcare institution during a period of one year, regardless of the diagnosis of the patients. Thus, our collection process ensures that the dataset is representative of a real clinical scenario, with patients having different risks of developing lung cancer, and that the data mimics the incidence of lung cancer in the population. We also collect the clinical report associated with each of the CT scans to have an integral understanding of the context regarding the medical history of the patient.

For each of the patients in the LUCAS dataset, we have the latest CT scan that was performed on the patient as well as the clinical report in which the physician explains the findings related to that CT scan and states if the patient has cancer or not. We anonymize all of the CT scans and clinical reports according to the established standards. We process this information to create a multimodal framework that includes visual information as well as a set of relevant biomarkers that might indicate risk factors.

2.1 Visual Information

The LUCAS dataset contains 830 low-dose chest CT scans from patients in a real-life setting. 72 of these patients are diagnosed with cancer by an expert physician. Nonetheless, most of these patients have a respiratory disease or are at high risk of developing lung cancer. The diversity in patient's diagnosis makes identifying visual patterns a complex task. We select 20% of the patients for the testing set and ensure that both groups share the same proportion of patients with and without cancer.

2.2 Biomarkers

For each patient in the LUCAS dataset, we have a clinical report associated with the patient's last CT scan. However, the information in these reports varies according to the level of detail registered by the physician. For this reason, we turn the clinical reports into sets of structured information to facilitate automated interpretation. We translate the reports into biomarkers that include relevant information for lung cancer diagnosis. Aided by expert physicians, we select characteristics that are transversal to every clinical report.

---

[1] https://github.com/BCV-Uniandes/LUCAS.

Table 1. Categories of biomarkers in the LUCAS Dataset

Category	Biomarkers
Cancer related factors	Cancer history, presence of pulmonary nodules, pulmonary nodule characteristics, presence of pulmonary masses, characteristics of pulmonary masses.
Clinical history	Respiratory history, previous CT scans, adenomegaly, thoracic pain, pleural effusion.
Visual analysis	Granulomas, pulmonary parenchymal consolidation, presence of tree-in-bud, opacities

The first category includes factors that are directly correlated with lung cancer such as medical history of cancer or factors that are a direct consequence of lung cancer like presence of pulmonary nodules. The second category is biomarkers regarding the patient's clinical history. The last category corresponds to visual analysis biomarkers. This category is composed by visual aspects of the CT scan that were highlighted by the physician in the report. Table 1 shows a description of the biomarkers in each category.

2.3 Task

For this dataset, we propose to study the screening task in which the distribution of the data resembles the real-life class imbalance. In this case, we have 8% of positive samples in the entire dataset. Additionally, as the evaluation metric, we propose to study this problem with the maximal F-score on the Precision-Recall curve [13,15].

3 Baseline Approach

3.1 Image Pre-processing

Variations in the voxel spacing of data may affect CNNs understanding of the images. For this reason, we resample the volumes to the median voxel spacing of the entire dataset using spline interpolation of third degree. In addition, the images are cropped along the depth dimension to include only the lungs of the patients. Finally, we perform a z-score normalization based on the statistics calculated for lung nodules in the Medical Segmentation Decathlon (MSD) [21] task for lung nodule segmentation.

3.2 Method

We train three methods to predict the probability of cancer for each patient. The models vary according to the modality of information used as input for

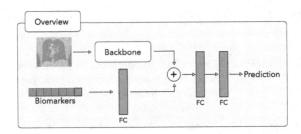

Fig. 1. Overview of the multimodal classification method. Our model extracts features from the diagnostic image and clinical report of a patient. Afterward, two fully connected (FC) layers combine the information to predict the probability of cancer.

the classification task. We test the effectiveness of using only visual information from the CT scans, using the biomarkers obtained from the clinical reports, and combining multimodal information.

Image-Based Approach: we use a classification network with a backbone pre-trained for lung nodule segmentation using the lung dataset from the MSD, and two fully connected layers to obtain the probability of cancer. The backbone has five stages with depthwise separable convolutions to reduce the computational cost derived from processing 3D images, and a strided convolution to reduce the image size. The number of feature maps is set to 32 in the first stage and is doubled after every dimensionality reduction, making sure that the maximum number of feature maps is 512. Also, to alleviate the issues derived from using small batches, we use instance normalization instead of the standard batch normalization.

Biomarkers: in this setting we use a multilayer perceptron to learn the relation between biomarkers and the relative importance of each one of them. The model contains a fully connected layer to encode the inputs followed by the same classification layers used for image classification.

Multimodal Approach: for the final model, we integrate both approaches for the individual modalities to obtain visual features from the CT scans and relevant information from the clinical reports. We perform the encoding stage for images and biomarkers in parallel and concatenate the resulting features to learn a joint representation using fully connected layers. Finally, the patient is classified according to the predicted probability of having cancer. An overview of the proposed pipeline is shown in Fig. 1

3.3   Training Details

We train our model for 40 epochs using Adam optimizer with weight decay of $1e - 5$ and an initial learning rate of $1e - 3$. The learning rate is reduced by a factor of 0.1 if the validation loss has not decreased in the previous 10 epochs.

Taking into account the large imbalance in the dataset, during training we assign higher probabilities of being selected to patients with cancer. By doing

Table 2. Results of the three variants of our baseline algorithm on the LUCAS Dataset.

Metric	Images	Reports	Multimodal data
ROC	0.513	0.674	**0.712**
F score	0.207	0.162	**0.250**

so, only a fraction of the negative samples is randomly selected every epoch, resulting in a natural data augmentation strategy.

4 Results

In order to assess quantitatively the difficulty of the lung cancer detection task on the LUCAS Dataset, as well as the relative importance of the different information modalities, we evaluate the three variants of our baseline algorithm on the test set.

Table 2 presents the results for both the ROC-AUC, the metric used in previous lung cancer datasets, and the maximal F-score on the Precision-Recall curve, the performance measure we adopt in this paper. The difference in absolute scores for the two metrics highlights the more stringent nature of the F-score and its appropriateness for detection tasks, as, in contrast to the ROC-AUC, it does not take into account true negatives in the computation. However, in both metrics, our multimodal baseline clearly outperforms the two versions of the system with only one modality. This result indicates that the two modalities provide complementary information and that our method is capable of taking benefit for improved detection. However, a closer look at the scores reveals that the maximal F-score is only 25% for the combined system, suggesting that lung cancer detection in the realistic setting of the LUCAS Dataset is still a very challenging problem, even in the presence of multimodal biomarkers.

In order to gain further insights on the results, we make use of the Toolkit for analyzing and visualizing challenge results [23] an evaluation framework that was designed to measure statistical significance of performance among different algorithms on biomedical machine learning challenges, and that was used to analyze the EndoVis 2019 Challenge results. Since the toolbox was created for a setting in which a performance metric such as the Dice Index is used to score softly algorithmic results rather than with binary detection labels, we use the detection probability as a score for positive instances and its complement for negative instances.

The main plots from the significance analysis are reported in Figs. 2 and 3. Figure 2 shows different possible rankings for the three algorithms, all based on the same individual scores. We can observe that the multimodal baseline is consistently ranked first, while the versions with a single modality can switch places depending on the specific ranking mechanism. This result underscores again the complementary nature of visual and clinical data, as well as the appropriateness of our multimodal system for leveraging it. It is also consistent with the apparent discrepancy in the ranking of individual modalities with the two metrics in Table 2. Figure 3 presents the dot-and-box plots for the individual scores, revealing a much tighter distribution for the multimodal system, and hence providing supporting evidence for the statistical significance of our results.

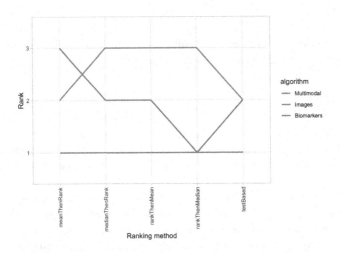

Fig. 2. Ranking robustness on three variants of our algorithm for the LUCAS Dataset. Using multimodal information proves to be more robust under every ranking.

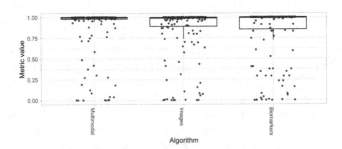

Fig. 3. Dot-and-box plot for the three variants of our algorithm on the LUCAS Dataset.

5   Conclusions

We present the LUCAS Dataset, the first experimental testbed for lung cancer detection with multimodal biomarkers. In addition to low-dose CT scans of hundreds of patients in a realistic clinical screening setup, we provide key clinical data from anonymized records and radiology reports to enrich the analysis. We develop a multimodal deep neural network as a strong baseline for future reference, and we show empirically the complementary nature of visual and clinical data for lung cancer detection. We hope that the availability of our experimental framework will enable the development of new generations of multimodal techniques and the exploration of new ideas for early lung cancer diagnosis.

Acknowledgments. This project was partially funded by the Google Latin America Research Awards (LARA) 2019.

References

1. Abbosh, C., et al.: Phylogenetic ctDNA analysis depicts early-stage lung cancer evolution. Nature **545**(7655), 446–451 (2017)
2. Alilou, M., Kovalev, V., Snezhko, E., Taimouri, V.: A comprehensive framework for automatic detection of pulmonary nodules in lung CT images. Image Anal. Stereol. **33**, 13 (2014). https://doi.org/10.5566/ias.v33.p13-27
3. Arbelaez, P., Maire, M., Fowlkes, C., Malik, J.: Contour detection and hierarchical image segmentation. IEEE Trans. Pattern Anal. Mach. Intell. **33**(5), 898–916 (2010)
4. Ardila, D., et al.: Endto-end lung cancer screening with three-dimensional deep learning on low-dose chest computed tomography. Nat. Med. **25**(6), 954–961 (2019). https://doi.org/10.1038/s41591-019-0447-x
5. Armato, S.G., et al.: LUNGx challenge for computerized lung nodule classification. J. Med. Imaging **3**(4), 044506 (2016). https://doi.org/10.1117/1.jmi.3.4.044506
6. Armato III, S.G., et al.: The lung image database consortium (LIDC) and image database resource initiative (IDRI): a completed reference database of lung nodules on CT scans. Med. Phys. **38**(2), 915–931 (2011)
7. Astaraki, M., Toma-Dasu, I., Smedby, Ö., Wang, C.: Normal appearance autoencoder for lung cancer detection and segmentation. In: Shen, D., et al. (eds.) MICCAI 2019. LNCS, vol. 11769, pp. 249–256. Springer, Cham (2019). https://doi.org/10.1007/978-3-030-32226-7_28
8. AstraZeneca, I.: Summary of product characteristics (2007)
9. Balagurunathan, Y., et al.: Isbi 2018 - lung nodule malignancy prediction challenge (2018). http://isbichallenges.cloudapp.net/competitions/15
10. Beckles, M.A., Spiro, S.G., Colice, G.L., Rudd, R.M.: Initial evaluation of the patient with lung cancer: symptoms, signs, laboratory tests, and paraneoplastic syndromes. Chest **123**(1), 97S–104S (2003)
11. Bray, F., Ferlay, J., Soerjomataram, I., Siegel, R.L., Torre, L.A., Jemal, A.: Global cancer statistics 2018: GLOBOCAN estimates of incidence and mortality worldwide for 36 cancers in 185 countries. CA: Cancer J. Clin. **68**(6), 394–424 (2018). https://doi.org/10.3322/caac.21492
12. del Ciello, A., Franchi, P., Contegiacomo, A., Cicchetti, G., Bonomo, L., Larici, A.R.: Missed lung cancer: when, where, and why? Diagn. Interv. Radiol. **23**(2), 118–126 (2017). https://doi.org/10.5152/dir.2016.16187
13. Flach, P., Kull, M.: Precision-recall-gain curves: PR analysis done right. In: Advances in Neural Information Processing Systems, pp. 838–846 (2015)

14. Hansell, D.M., Bankier, A.A., MacMahon, H., McLoud, T.C., Müller, N.L., Remy, J.: Fleischner society: glossary of terms for thoracic imaging. Radiology **246**(3), 697–722 (2008). https://doi.org/10.1148/radiol.2462070712

15. Hariharan, B., Arbeláez, P., Bourdev, L., Maji, S., Malik, J.: Semantic contours from inverse detectors. In: 2011 International Conference on Computer Vision, pp. 991–998. IEEE (2011)

16. Jacobs, C., Setio, A.A.A., Traverso, A., van Ginneken, B.: Luna16 (2016). https://luna16.grand-challenge.org

17. Kaggle: Data science bowl 2017 (2017). https://www.kaggle.com/c/data-science-bowl-2017

18. Memon, W., et al.: Can computer assisted diagnosis (CAD) be used as a screening tool in the detection of pulmonary nodules when using 64-slice multidetector computed tomography? Int. J. Gen. Med. **4**, 815 (2011). https://doi.org/10.2147/ijgm.s26127

19. Pedrosa, J., Ferreira, C., Aresta, G.: Grand challenge on automatic lung cancer patient manager (2020). https://lndb.grand-challenge.org/

20. Pérez, G., Arbeláez, P.: Lung cancer prediction (2018). https://biomedicalcomputervision.uniandes.edu.co/index.php/research?id=33

21. Simpson, A.L., et al.: A large annotated medical image dataset for the development and evaluation of segmentation algorithms. CoRR abs/1902.09063 http://arxiv.org/abs/1902.09063 (2019)

22. Way, T., et al.: Computer-aided diagnosis of lung nodules on CT scans. Acad. Radiol. **17**(3), 323–332 (2010). https://doi.org/10.1016/j.acra.2009.10.016

23. Wiesenfarth, M., Reinke, A., Landman, A.L., Cardoso, M., Maier-Hein, L., Kopp-Schneider, A.: Methods and open-source toolkit for analyzing and visualizing challenge results. arXiv preprint arXiv:1910.05121 (2019)

24. Wong, M.C., Lao, X.Q., Ho, K.F., Goggins, W.B., Shelly, L.: Incidence and mortality of lung cancer: global trends and association with socioeconomic status. Sci. Rep. **7**(1), 1–9 (2017)

Automatic Breast Lesion Classification by Joint Neural Analysis of Mammography and Ultrasound

Gavriel Habib[1]([✉]), Nahum Kiryati[2], Miri Sklair-Levy[3], Anat Shalmon[3],
Osnat Halshtok Neiman[3], Renata Faermann Weidenfeld[3], Yael Yagil[3],
Eli Konen[3], and Arnaldo Mayer[3]

[1] School of Electrical Engineering, Tel-Aviv University, Tel Aviv-Yafo, Israel
gavrielhabib@mail.tau.ac.il
[2] The Manuel and Raquel Klachky Chair of Image Processing,
School of Electrical Engineering, Tel-Aviv University, Tel Aviv-Yafo, Israel
[3] Diagnostic Imaging, Sheba Medical Center, Affiliated to the Sackler School
of Medicine, Tel-Aviv University, Tel Aviv-Yafo, Israel

Abstract. Mammography and ultrasound are extensively used by radiologists as complementary modalities to achieve better performance in breast cancer diagnosis. However, existing computer-aided diagnosis (CAD) systems for the breast are generally based on a single modality. In this work, we propose a deep-learning based method for classifying breast cancer lesions from their respective mammography and ultrasound images. We present various approaches and show a consistent improvement in performance when utilizing both modalities. The proposed approach is based on a GoogleNet architecture, fine-tuned for our data in two training steps. First, a distinct neural network is trained separately for each modality, generating high-level features. Then, the aggregated features originating from each modality are used to train a multimodal network to provide the final classification. In quantitative experiments, the proposed approach achieves an AUC of 0.94, outperforming state-of-the-art models trained over a single modality. Moreover, it performs similarly to an average radiologist, surpassing two out of four radiologists participating in a reader study. The promising results suggest that the proposed method may become a valuable decision support tool for breast radiologists.

Keywords: Deep learning · Mammography · Ultrasound

1 Introduction

Breast cancer is the second most common type of cancer among American women after skin cancer. According to the American Cancer Society estimations, 268,600 invasive breast cancer cases have been diagnosed in 2019, leading

© Springer Nature Switzerland AG 2020
T. Syeda-Mahmood et al. (Eds.): ML-CDS 2020/CLIP 2020, LNCS 12445, pp. 125–135, 2020.
https://doi.org/10.1007/978-3-030-60946-7_13

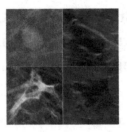

Fig. 1. Benign (top) and Malignant (bottom) lesions from our dataset. Malignant lesions tend to have less strict boundaries in both mammography (left) and ultrasound (right) screenings.

to 41,760 deaths. However, early detection may save lives as it enables better treatment options.

Mammography-based screening is the most widely used approach for breast cancer detection, with proven mortality reduction and early disease treatment benefits [1]. However, it suffers from poor lesion visibility in dense breasts [2]. To improve sensitivity in dense breasts, contrast-enhanced spectral mammography (CESM) has been developed. CESM is based on the subtraction of low and high energy images, acquired following the injection of a contrast agent [3]. Although CESM reaches MRI levels of lesion visibility for dense breasts [31], the technique is still in the early adoption phase.

Ultrasound imaging has proven to be a valuable tool in dense breasts, increasing cancer detection sensitivity by 17% [4]. Nevertheless, breast ultrasound may miss solid tumors that are easily detected with mammography. Devolli-Disha et al. [5] showed that ultrasound had a higher sensitivity (69.2%) than mammography (15.4%) in women younger than 40 years, whereas mammography (78.7%) beats ultrasound (63.9%) in women older than 60 years. Due to its benefits and disadvantages, radiologists suggest using breast ultrasound as a complementary screening test to mammography [6,7].

Classification of breast lesions is a challenging task for the radiologist. Malignant and benign lesions can be differentiated by their shape, boundary and texture. For example, malignant lesions may have irregular and not well defined boundaries as they have the ability to spread (see Fig. 1). Nevertheless, in many cases radiologists cannot classify the lesion and the patient is referred for a biopsy which is a stressful and expensive process. Given that 65%–85% of the biopsies turns out to be benign [8], there is a clear need for tools that will help radiologists reduce benign biopsies.

2   Related Work

In recent years, deep learning techniques have been providing significant improvements in various medical imaging tasks, such as tumor detection and classification, image denoising and registration. In the field of breast cancer classification, existing methods are based mainly on mammograms [9,17–19], ultrasound [10,20], MRI [21,22] or histopathology images [23].

To deal with the limited amount of data, Chougrada et al. [9] used transfer learning over ImageNet and achieved state-of-the-art results over public mammography datasets. Cheng et al. [10] performed a semi-supervised learning approach over a large breast ultrasound dataset with only few annotated images. Wu et al. [19] synthesized mammogram lesions using class-conditional GAN and used them as additional training data instead of basic augmentations.

Emphasizing the importance of lesions' context, Wu et al. [17] trained a deep multi-view CNN over a large private mammogram dataset. They used a breast-level model to create heatmaps that represent suspected areas, and a patch-level model to locally predict the presence of malignant or benign findings. Shen et al. [18] combined coarse and fine details using an attention mechanism to select informative patches for classification.

Common breast imaging modalities were also combined with additional data from other domains. Byra et al. [11] used the Nakagami parameter maps created from breast ultrasound images to train a CNN from scratch. Perek et al. [12] integrated CESM images with features of BIRADS [13], a textual radiological lexicon for breast lesions, as inputs to a classifier.

Most of previous studies utilized only a single modality, while some combined different types of breast images. Hadad et al. [14] classified MRI breast lesions using fine tuning of a network pre-trained on mammography images instead of natural images. Regarding mammography with ultrasound, Cong et al. [15] separately trained three base classifiers (SVM, KNN and Naive Bayes) for each modality, integrated some of them by a selective ensemble method and obtained the final prediction by majority vote. Shaikh et al. [16] proposed a learning-using-privileged-information approach, i.e. utilizing both modalities for training, but avoiding one during test time. These papers suggested the potential of cross-modal learning.

In this paper, we propose a novel deep-learning method for the classification of breast lesion, using both mammography and ultrasound images of the lesion. To the best of our knowledge, it is the first reported attempt to combine these very different imaging modalities by fusing high-level perceptual representations for lesion classification. We use a unique dataset consisting of matched mammography and ultrasound lesions, acquired at our institution. The proposed methods are evaluated using a leave one out scheme, demonstrating significant improvement in AUC (area under curve) when features extracted from both modalities are combined into a single multi-modality classifier, in comparison to single modality classification using only mammography or ultrasound.

3   Method

3.1   Dataset

Although combining mammography and ultrasound imaging for breast cancer screenings is a common practice, to the best of our knowledge there are no public datasets containing corresponding lesions from both modalities. Therefore, we created our own retrospective dataset of 153 biopsy-proven lesions, consisting of 73 malignant and 80 benign cases. For each lesion, corresponding mammography and ultrasound images were contoured by an expert breast radiologist, with a biopsy proven labelling. Figure 2 demonstrates a sample from the dataset.

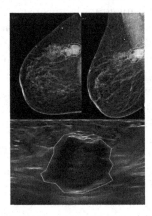

Fig. 2. Matched malignant lesion contouring in both modalities.

3.2 Model Architecture

Single Modality Networks. Two convolutional neural networks (CNNs), one for each modality, were trained to tell apart malignant and benign lesions. The contoured lesions were cropped into image patches and submitted to geometric transformations (translation, rotation, flipping) to augment the dataset and generate additional inputs.

We experimented with two different architectures: (1) Basic CNN with ReLU activation maps, max pooling and fully connected layers that was trained from scratch (Fig. 3); (2) GoogleNet [24] previously trained over ImageNet.

Multimodal Network. Figure 4 presents the multimodal fully connected network, consisting of 7 layers. High-level perceptual descriptors of matched lesions were extracted from both trained single-modality networks and combined by concatenation. The concatenated vector is then used as an input for the multimodal network, which eventually provides the final malignancy probability of the input lesion.

3.3 Implementation Details

Loss Function. Both single and multi modal classifiers were trained using the same loss function in each experiment. To enrich diversity, we experimented

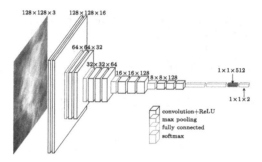

Fig. 3. Proposed single modality basic CNN architecture. The input is a cropped lesion and the output is the softmax malignancy probability. A 512 dimensional vector ("descriptor") is the last layer before the output layer.

Fig. 4. Multimodality fully connected network architecture. The input is a concatenation of correspondence lesion descriptors extracted from mammography and ultrasound CNNs. The output is the softmax malignancy probability.

with two loss functions: (1) BCE - Binary Cross Entropy loss; (2) LMCL - Large Margin Cosine Loss [25] which is commonly used in Face Recognition tasks. LMCL defines a decision margin in the cosine space and learns discriminative features by maximizing inter-class and minimizing intra-class cosine margin.

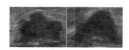

Fig. 5. Same lesion captured in different views in ultrasound screening.

Training Method. Lesion patches from different modalities can be utilized for classification in different manners. Unfortunately, image registration is almost impossible because of the difference in mammography and ultrasound imaging techniques. Moreover, even ultrasound images of the same lesion are highly different, as the images are captured in various views and the breast is easily deformed by the mechanical pressure applied by the transducer (see Fig. 5).

Therefore, we make use of the coupled mammography-ultrasound lesions by combining them in the feature space instead of image space. For completeness, we show two different training methods: (1) We first train each single modality network separately, then combine high-level feature data from both networks and feed it as an input for training the multimodal network; (2) End-to-end training, illustrated in Fig. 6, in which we train all three networks (two single modality networks and one fully connected after feature combination) concurrently. The loss function is the sum of the losses from each of the three networks. In this approach, the performance of each network is tied to the other two's.

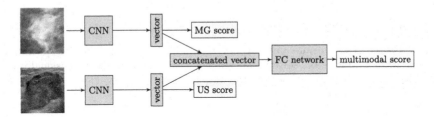

Fig. 6. End-to-end training method: all three classifiers are trained at the same time, while the loss is the sum of all three losses.

4   Experiments

4.1   Leave-One-Out

Given 153 mammography-ultrasound lesion pairs, we randomly selected 120 fixed pairs for the leave-one-out experiments, benign and malignant being equally distributed. The remaining 33 lesions were held out as validation set for hyper-parameter tuning.

We ran 8 different experiments, one for each combination of previously mentioned configurations: training method, model architecture and loss function. All the experiments were performed using the "leave-one-out" methodology. In each phase, 119 out of 120 lesions were used for training and a single lesion, different in every phase, was used for testing. Finally, the test lesion obtained three scores, one for each modality and one combined, representing the average malignancy probabilities of all its appearances in the dataset.

Results were evaluated by means of AUC (area under the ROC curve), calculated from all test scores. As can be seen in Table 1, combining mammography and ultrasound features improved results in most experiments. In fact, only in one experiment results deteriorated due to the combination of modalities. Clearly, using transfer learning outperforms training from scratch, likely because of the small dataset size. Moreover, two steps of training (each modality first and combined descriptors afterwards) achieve better results than an end-to-end training. No significant difference is observed in the performance of the tested loss functions.

Comparison to State-of-the-Art Models. Our method is based on combining mammography with ultrasound. However, as previous authors haven't discussed their exact models' design [15], we report the results of two baselines, each using a single modality, and compare them to our single modality networks. As each model was trained over a different dataset, it may be confusing and even meaningless to directly compare reported results. Therefore, for qualitative assessment of our model, we trained these models over our own dataset.

Mammography. The patch-level network proposed by [17] was trained on our mammography dataset. Based on DenseNet121 [27] and transfer learning, it achieved an AUC of 0.86 - better than our trained from scratch CNN (0.76), but inferior to GoogleNet (0.89).

Ultrasound. Training VGG16 model [29] previously trained over ImageNet, on our ultrasound dataset, as suggested by [28], yielded AUC of 0.81. It is worth mentioning that the reported AUC on the original dataset of Hijab et al. was much higher (0.97), which may suggest that their method is sensitive to the specific training data used. The obtained AUC is better than our trained from scratch CNN (0.75), but inferior to GoogleNet (0.88).

Table 1. AUC results of all experiments, reported on test set. Scores order is as follows: mammography/ultrasound/combined.

Training method	Loss function	Initialization	
		From scratch (CNN)	Transfer learning (GoogleNet)
Separate	BCE	0.76/0.75/**0.79**	0.89/0.88/**0.94**
	LMCL	0.73/0.79/**0.82**	0.88/0.87/**0.89**
End-to-end	BCE	0.74/0.78/**0.79**	**0.82**/0.81/**0.82**
	LMCL	0.72/0.78/**0.79**	**0.84**/0.80/0.78

4.2 Reader Study

To compare the proposed method with human radiologists, we performed a simplified reader study with 4 experienced radiologists. 120 pairs of corresponding mammography-ultrasound lesion images, taken from the biopsy-proven leave-one-out experiment dataset, were visually assigned a malignancy rate (from 0 to 10) by each of the participating radiologists separately. The AUCs achieved by the readers were: 0.931, 0.938, 0.967 and 0.979, compared to 0.942 for our best model. ROC curves are shown in Fig. 7. These results suggest that the proposed model performed similarly to an average radiologist.

4.3   Model Insight

Is the model paying attention to the same attributes as radiologists when predicting whether a lesion is malignant? To gain insight about this question, we applied the Grad-CAM algorithm [26] to our best trained GoogleNet. Grad-CAM produces a gradient-based heat-map that highlights input parts that most influenced the output prediction.

In Fig. 8, we present several mammography and ultrasound malignant examples from the training set, with their Grad-CAM computations. "Hotter" areas indicate attended regions. We observe that for malignant lesions, the model appears to rely significantly (hot colors in heat map) on the lesion boundaries, especially where irregular features are encountered, in agreement with the radiologist diagnostic methodology.

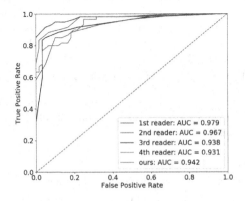

Fig. 7. ROC curves of our model and each reader.

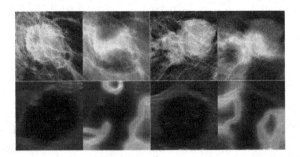

Fig. 8. Examples of malignant lesions from the training set with their Grad-CAM visualizations (top-mammography, bottom-ultrasound).

5 Conclusion

We propose a deep-learning method for the classification of breast lesions that combines mammography and ultrasound input images. We show that by combining high-level perceptual features from both modalities, the classification performance is improved. Furthermore, the proposed method is shown to perform similarly to an average radiologist, surpassing two out of four radiologists participating in a reader study. The promising results suggest the proposed method may become a valuable decision support tool for multimodal classification of breast lesions. In future research, further validation on a larger dataset should be performed. The proposed method may be generalized by incorporating additional imaging modalities, such as breast MRI, as well as medical background information of the patient [30].

References

1. Heywang-Köbrunner, S.H., Hacker, A., Sedlacek, S.: Advantages and disadvantages of mammography screening. Breast Care **6**(3), 199–207 (2011)
2. Carney, P.A., et al.: Individual and combined effects of age, breast density, and hormone replacement therapy use on the accuracy of screening mammography. Ann. Intern. Med. **138**(3), 168–75 (2003)
3. Lobbes, M.B.I., Smidt, M.L., Houwers, J., Tjan-Heijnen, V.C., Wildberger, J.E.: Contrast enhanced mammography: techniques, current results, and potential indications. Clin. Radiol. **68**(9), 935–944 (2013)
4. Kolb, T.M., et al.: Occult cancer in women with dense breasts: detection with screening US-diagnostic yield and tumor characteristics. Radiology **207**(1), 191–199 (1998)
5. Devolli-Disha, E., Manxhuka-Kerliu, S., Ymeri, H., Kutllovci, A.: Comparative accuracy of mammography and ultrasound in women with breast symptoms according to age and breast density. Bosn. J. Basic Med. Sci. **9**, 131–136 (2009)
6. Kelly, K.M., Dean, J., Lee, S.-J., Comulada, W.S.: Breast cancer detection: radiologists' performance using mammography with and without automated whole-breast ultrasound. Eur. Radiol. **20**(11), 2557–2564 (2010)
7. Skaane, P., Gullien, R., Eben, E.B., Sandhaug, M., Schulz-Wendtland, R., Stoeblen, F.: Interpretation of automated breast ultrasound (ABUS) with and without knowledge of mammography: a reader performance study. Acta Radiol. **56**(4), 404–412 (2015)
8. Jesneck, J.L., Lo, J.Y., Baker, J.A.: Breast mass lesions: computer-aided diagnosis models with mammographic and sonographic descriptors. Radiology **244**(2), 390 (2007)
9. Chougrada, H., Zouakia, H., Alheyane, O.: Deep convolutional neural networks for breast cancer screening. Comput. Methods Programs Biomed. **157**, 19–30 (2018)
10. Cheng, J.-Z., Ni, D., Chou, Y.-H., Qin, J., Tiu, C.-M., Chang, Y.-C., et al.: Computer-aided diagnosis with deep learning architecture: applications to breast lesions in US images and pulmonary nodules in CT scans. Sci. Rep. **6**, 24454 (2016)
11. Byra, M., Piotrzkowska-Wroblewska, H., Dobruch-Sobczak, K., Nowicki, A.: Combining Nakagami imaging and convolutional neural network for breast lesion classification. Paper Presented at the IEEE International Ultrasonics Symposium, IUS (2017)

12. Perek, S., Kiryati, N., Zimmerman-Moreno, G., Sklair-Levy, M., Konen, E., Mayer, A.: Classification of contrast-enhanced spectral mammography (CESM) images. Int. J. Comput. Assist. Radiol. Surg. **14**, 249–257 (2019)
13. American College of Radiology: ACR BI-RADS Atlas 5th edn., pp. 125–143 (2013)
14. Hadad, O., Bakalo, R., Ben-Ar, R., Hashoul, S., Amit, G.: Classification of breast lesions using cross-modal deep learning. In: IEEE International Symposium on Biomedical Imaging (ISBI) (2017)
15. Cong, J., Wei, B., He, Y., Yin, Y., Zheng, Y.: A selective ensemble classification method combining mammography images with ultrasound images for breast cancer diagnosis. Comput. Math. Methods Med. **2017**, 1–7 (2017)
16. Shaikh, T.A., Ali, R., Beg, M.M.S.: Transfer learning privileged information fuels CAD diagnosis of breast cancer. Mach. Vis. Appl. **31**(1), 9 (2020)
17. Wu, N., et al.: Deep neural networks improve radiologists' performance in breast cancer screening. IEEE Trans. Med. Imaging **39**, 1184–1194 (2019)
18. Shen, Y., et al.: Globally-aware multiple instance classifier for breast cancer screening. In: Suk, H.-I., Liu, M., Yan, P., Lian, C. (eds.) MLMI 2019. LNCS, vol. 11861, pp. 18–26. Springer, Cham (2019). https://doi.org/10.1007/978-3-030-32692-0_3
19. Wu, E., Wu, K., Cox, D., Lotter, W.: Conditional infilling GANs for data augmentation in mammogram classification. In: Stoyanov, D., et al. (eds.) RAMBO/BIA/TIA-2018. LNCS, vol. 11040, pp. 98–106. Springer, Cham (2018). https://doi.org/10.1007/978-3-030-00946-5_11
20. Han, S., et al.: A deep learning framework for supporting the classification of breast lesions in ultrasound images. Phys. Med. Biol. **62**(19), 7714 (2017)
21. Bevilacqua, V., Brunetti, A., Triggiani, M., Magaletti, D., Telegrafo, M., Moschetta, M.: An optimized feed-forward artificial neural network topology to support radiologists in breast lesions classification. In: Proceedings of the 2016 on Genetic and Evolutionary Computation Conference Companion, pp. 1385–1392 (2016)
22. Amit, G., et al.: Classification of breast MRI lesions using small-size training sets: comparison of deep learning approaches. In: Medical Imaging 2017: Computer-Aided Diagnosis, vol. 10134. International Society for Optics and Photonics (2017)
23. Araújo, T., et al.: Classification of breast cancer histology images using convolutional neural networks. PloS One **12**(6), e0177544 (2017)
24. Szegedy, C., et al.: Going deeper with convolutions. In: Proceedings of the IEEE Conference on Computer Vision and Pattern Recognition, pp. 1–9 (2015)
25. Wang, H., et al.: Cosface: large margin cosine loss for deep face recognition. In: Proceedings of the IEEE Conference on Computer Vision and Pattern Recognition, pp. 5265–5274 (2018)
26. Selvaraju, R.R., Cogswell, M., Das, A., Vedantam, R., Parikh, D., Batra, D.: Grad-CAM: visual explanations from deep networks via gradient-based localization. In: Proceedings of the IEEE International Conference on Computer Vision, pp. 618–626 (2017)
27. Huang, G., Liu, Z., Van Der Maaten, L., Weinberger, K.Q.: Densely connected convolutional networks. In: Proceedings of the IEEE Conference on Computer Vision and Pattern Recognition, pp. 4700–4708 (2017)
28. Hijab, A., Rushdi, M.A., Gomaa, M.M., Eldeib, A.: Breast cancer classification in ultrasound images using transfer learning. In: 2019 Fifth International Conference on Advances in Biomedical Engineering (ICABME), pp. 1–4. IEEE (2019)
29. Simonyan, K., Andrew, Z.: Very deep convolutional networks for large-scale image recognition. arXiv preprint arXiv:1409.1556 (2014)

30. Yala, A., et al.: Using machine learning to parse breast pathology reports. Breast Cancer Res. Treat. **161**(2), 203–211 (2017)
31. Li, L., et al.: Contrast-enhanced spectral mammography (CESM) versus breast magnetic resonance imaging (MRI): a retrospective comparison in 66 breast lesions. Diagn. Interv. Imaging **98**(2), 113–123 (2017)

Author Index

Aktar, Mumu 24
Arbeláez, Pablo 115
Arth, Clemens 64

Bakstein, Eduard 34
Bao, Shunxing 13
Berg, Astrid 97
Beyer, Thomas 97
Bianchi, Jonas 44
Bühler, Katja 97

Campe, Gord von 75
Carr, John J. 13
Castillo, Angela 115
Castillo, Aron Aliaga-Del 44
Cevidanes, Lucia 44
Chaves Junior, Cauby Maia 44

Daza, Laura 115
De Oliveira Ruellas, Antonio Carlos 44

Egger, Jan 64, 75
Escobar, María 115
Evangelista, Karine 44

Faermann Weidenfeld, Renata 125
Fillion-Robin, Jean-Christophe 44
Fleck, Philipp 64

Gao, Riqiang 13
Garib, Daniela Gamba 44
Gilmore, Greydon 34
Gonçalves, João Roberto 44
Gsaxner, Christina 64, 75
Gurgel, Marcela Lima 44
Guy, Pierre 3

Habib, Gavriel 125
Hager, Gregory D. 54
Halshtok Neiman, Osnat 125
Henriques, Jose Fernando Castanha 44
Hibbard, Lyndon 85
Hoctor, James 44

Hodgson, Antony J. 3
Huo, Yuankai 13

Ishii, Masaru 54
Iwasaki, Laura R. 44

Janson, Guilherme 44

Karner, Florian 64
Kersten-Oertel, Marta 24
Kiryati, Nahum 125
Konen, Eli 125

Landman, Bennett A. 13
Lee, Ho Hin 13
Lefaivre, Kelly A. 3
Lenis, Dimitrios 97
Li, David Z. 54
Li, Jianning 64, 75
Li, TengFei 44

Major, David 97
Massaro, Camila 44
Mayer, Arnaldo 125

Neubauer, Theresa 97
Nickel, Jeffrey C. 44
Novak, Daniel 34

Pandey, Prashant U. 3
Paniagua, Beatriz 44
Pepe, Antonio 64, 75
Pinzón, Bibiana 115
Prietro, Juan C. 44

Rivaz, Hassan 24

Saponjski, Jelena 97
Shalmon, Anat 125
Sinha, Ayushi 54
Sklair-Levy, Miri 125
Spann, Ashley 13
Styner, Martin 44

Tampieri, Donatella   24
Tang, Yucheng   13
Taylor, Russell H.   54
Terry, James G.   13

Valencia, Sergio   115
van Sonsbeek, Tom   106
Varga, Igor   34
Vilanova, Lorena   44

Wallner, Jürgen   64
Wells, Quinn Stanton   13
Wimmer, Maria   97
Worring, Marcel   106

Xiao, Yiming   24

Yagil, Yael   125
Yatabe, Marília   44

Printed in the United States
By Bookmasters